SECOND EDITION

Standard Grade
PHYSICS

Drew McCormick
Arthur Baillie

Hodder Gibson

A MEMBER OF THE HODDER HEADLINE GROUP

Acknowledgements

The publishers would like to thank the following individuals, institutions and companies for permission to reproduce photographs in this book. Every effort has been made to trace ownership of copyright. The publishers would be happy to make arrangements with any copyright holder whom it has not been possible to contact:

Andrew Lambert (30, 56, 58, 72 bottom, 147 bottom); AP Photos (85); Arthur Baillie (20, 36, 42, 43, 47, 54, 131 both); Corbis (39); Empics (148 bottom); Glasgow Science Centre/Keith Hunter (163); Hodder & Stoughton (10, 30 top, 168 bottom); John Birdsall Photography (135 top); Kodansha (211); Life File (135 bottom); PA Photos (22 right, 25, 88 left, 143 top); Phillips Consumer Electrics (18); Ronald Grant (17); Science Photo Library (2, 7 bottom, 9, 22 left, 23, 28, 57, 66 both, 70, 71 top, 73 both, 77, 80, 82, 89 top two, 90 all, 94, 96, 99, 100 both, 101, 105, 143 bottom, 147 top. 148 top, 149, 153 both, 154, 168 top, 169, 197 both, 198, 200 both, 202 all, 203, 206 both, 209 both, 213 both, 215 both); Sperry Marine Systems (7 top); Wellcome Trust (71 bottom three, 72 top and middle, 84, 87 both. 88 right, 89 bottom two, 102 both).

Orders: please contact Bookpoint Ltd, 130 Milton Park, Abingdon, Oxon OX14 4SB. Telephone: (44) 01235 827720, Fax: (44) 01235 400454. Lines are open from 9.00 – 5.00, Monday to Saturday, with a 24 hour message answering service.
You can also order through our website www.hoddereducation.co.uk

British Library Cataloguing in Publication Data
A catalogue record for this title is available from The British Library

Published by Hodder Gibson, 2a Christie Street, Paisley PA1 1NB.
Tel: 0141 848 1609; Fax: 0141 889 6315; email: hoddergibson@hodder.co.uk

ISBN -13: 978-0-340-84710-7

First edition published 1997
This edition published 2002
Impression number   10 9 8 7 6 5
Year   2007

ISBN -13: 978-0-340-84711-4

First edition published 1997
This edition published 2002
Impression number   10 9 8 7 6
Year   2007
With Answers

Printed in Italy for Hodder Gibson, an imprint of Hodder Education, an Hachette Livre UK Company, 2a Christie Street, Paisley PA1 1NB, Scotland, UK

CONTENTS

PREFACE

'All science is either physics or stamp collecting'
Lord Rutherford – Nobel Prize winner

'My goal is simple. It is the complete understanding of the universe, why it is and why it exists at all'

Stephen Hawking, Lucasian Professor of Mathematics, University of Cambridge and author of several books including 'A Brief History of Time' and 'Stephen Hawking's Universe'.

This book represents a review of the previously successful textbook, which we wrote in response to the growing demands for an interesting book, which reflected both the applications led approach, and the rigour and consistency to allow candidates to achieve a good understanding of the basic principles of physics.

Our aim is to increase the understanding of the main features of physics in the world and to excite that spark which will lead to a greater understanding and interest in the subject.

We have been able to use full colour in the book and to add a wider variety of relevant applications. The introduction of fascinating physics is not part of the course but includes new applications, particularly in the field of health physics. We have also taken account of the response of our students, and learning outcomes and summaries are inserted at the relevant points in the chapters.

We are grateful to many people for help in the production of this book. In particular we owe an invaluable debt to Roddy Glen for his painstaking work in editing the book. He has noticed omissions and corrected many problems in the draft manuscript. We are also grateful to many staff at Hodder and Stoughton, particularly to Elisabeth Tribe for commissioning a new edition and to Ruth Hughes and Charlotte Litt for their detailed work on the manuscript.

Andrew McCormick
Arthur Baillie

January 2002

CHAPTER ONE

Telecommunications

Brian was driving to the airport to meet some friends but he was unsure of his directions. He searched the internet and found the directions from his house to the airport. On his way he used a computer in his car to help give him immediate directions taking account of traffic. He used a GPS (Global Positioning System) computer on his car. The computer used a GPS to tell the car where it was in relation to a geostationary satellite. This then gave him directions about the route to follow. When he came close to the airport he stopped and used his mobile phone. This could connect him to the internet via the WAP (Wireless Application Protocol) system and allowed him to check the arrival times of flights. Such a system of navigation exists and replaces out of date maps and fixed line telephones.

We live at a time when communications are important as part of every day life. We use faxes and the telephone, mobile phones and the internet. In just over 100 years we have gone from the simple telephone to browsing the web. This chapter explains the ideas behind various forms of modern communications.

SECTION 1.1 Communication using waves

At the end of this section you should be able to:

1 Give an example which illustrates that the speed of sound in air is less than the speed of light in air, e.g. thunder and lightning.
2 Describe a method of measuring the speed of sound in air (using the relationship between distance, time and speed).
3 Carry out calculations involving the relationship between distance, time and speed in problems on sound transmission.
4 State that waves are one way of transmitting signals.
5 Use the following terms correctly in context: wave, frequency, wavelength, speed, energy (transfer), amplitude.
6 Carry out calculations involving the relationship between distance, time and speed in problems on water waves.
7 Carry out calculations involving the relationship between speed, wavelength and frequency for water and sound waves.

8 Explain the equivalence of $f \times \lambda$ and $\frac{d}{t}$

Communication

Communication is when information (a message) is successfully transmitted or sent and received. This chapter looks at communication in different ways. (figure 1.1). Table 1.1 lists some important events in the history of communications.

Figure 1.1 Modern communications.

Event	Date
Samuel Morse uses his code to send a message	1835
Cooke and Wheatstone use the electric telegraph on the railways	1837
James Clerk Maxwell from Edinburgh shows mathematically that electricity can be sent as waves	1864
Alexander Graham Bell develops the telephone	1876
Heinrich Hertz shows that electricity can be sent as waves	1888
Marconi sends the first wireless message across the English Channel	1899
First radio broadcast in the UK and USA	1920
John Logie Baird demonstrates television	1926
First video recorder	1958
Communications satellite launched called 'Telstar'	1962

Table 1.1 Significant events in the development of communications.

Telecommunication

The word 'tele' comes from the Greek word meaning 'at a distance'. Early communications used sound and light signals to send messages. We shall be looking at these two methods of communication. Light travels faster than sound. This can be seen during a thunderstorm when you see the lightning before you hear the thunder – unless the storm is overhead.

We can compare the speed of sound with the speed of light. For example, when a car is a few hundred metres away from a person who steps into the road to cross, the driver sounds his horn and at the same instant the lights are switched on by the passenger. The lights are seen by that person before the sound of the horn is heard. Often you will see an aircraft before you hear the sound of the engines.

This tells us that light travels faster than sound, but how fast does sound travel? To answer this question we need to measure the speed of sound.

Speed

Speed is defined as the distance travelled in one second by a moving object.

Example
June runs a distance of 200 m in 40 seconds.

$$\text{June travels 200 m in 40 s}$$

$$\text{June's Speed} = \frac{200}{40} = 5\,\text{m/s}$$

In our example 200 m was the distance travelled and 40 s was the time taken:

$$\text{Speed} = \frac{\text{distance travelled}}{\text{time taken}}$$

or $\quad v = \dfrac{d}{t} \qquad$ where v = speed in metres per second (m/s)
d = distance travelled in metres (m)
t = time taken in seconds (s)

Measuring the speed of sound

1 A pupil with a source of sound (e.g. a pair of cymbals) stands at one end of the field.
2 At the other end of the field is a pupil with a stop-watch (figure 1.2).
3 The length of the field is measured in metres.
4 The timekeeper starts the watch when she sees the cymbals coming together.
5 When the timekeeper hears the sound, the watch is stopped and the time noted.

The experiment is repeated and an average time calculated. Typical results might be:

Times noted = 0.65 s, 0.55 s, 0.58 s, 0.62 s
Average time = 0.6 s
Length of field = 200 m = distance travelled

$$\text{Speed of sound} = \frac{\text{distance travelled}}{\text{average time taken}} = \frac{200}{0.6} = 333\,\text{m/s}$$

Cymbals   Stop-watch

Figure 1.2 Measuring the speed of sound using cymbals and stop-watch.

There will always be some uncertainty in the measurement of time due to reaction time. This is the time taken to react to the event from seeing it happen. It is better to repeat the measurements of the times and calculate an average for the time taken.

To eliminate the reaction time problem which exists with the previous method, a more accurate method uses a computer and a timer interface to obtain the measurements of the time interval (figure 1.3):

1 The microphones are placed at least 1 metre apart.
2 The microphones are connected to the computer and the computer switched on.
3 The distance between the microphones is measured with a metre rule.
4 Hitting a metal plate with the hammer makes a loud sound.
5 When the sound reaches microphone X, the timer starts timing.
6 When the sound reaches microphone Y, the timer stops timing.

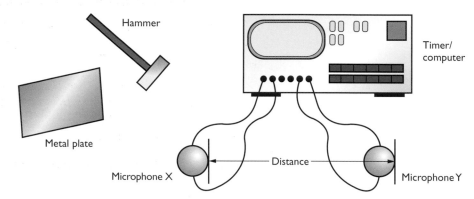

Figure 1.3 Measuring the speed of sound using a timer.

Typical results are:

$$\text{Distance between the microphones} = 1.0 \, \text{m}$$

$$\text{Time on the timer} = 3 \, \text{ms} = \frac{3}{1000}$$

$$= 0.003 \, \text{s (ms is a millisecond which is 1/1000 s)}$$

$$\text{Speed of sound} = \frac{\text{distance travelled}}{\text{time taken}} = \frac{1}{0.003}$$

$$\text{Value of speed of sound} = 330 \, \text{m/s}$$

Waves and wave patterns

Waves carry energy from one place to another and so can carry signals from one place to another. The water waves that roll onto a beach are transferring kinetic energy.

A typical wave pattern is shown in figure 1.4.

◆ The top part of the wave is called the *crest* and the bottom part is the *trough*.
◆ The line running through the middle of the wave pattern is called the *axis*.
◆ The distance from the axis to the top of the crest or the bottom of the trough is called the *amplitude*.
◆ The *wavelength* is the distance after which the pattern repeats itself.
◆ It is given the symbol λ (lambda) and is measured in metres or occasionally centimetres.

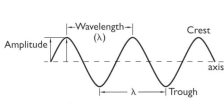

Figure 1.4 Wave characteristics.

$$\text{The frequency of the wave} = \frac{\text{number of waves produced}}{\text{time taken in seconds}}$$

It is measured in hertz (Hz).

While the speed of the wave can be found from the usual equation.

$$\text{Speed} = \frac{\text{distance travelled}}{\text{time taken}}$$

$$v = \frac{d}{t}$$ where v = speed in m/s
d = distance travelled in m
t = time taken in s

The speed of the wave can also be found from the following equation:

$$\text{Speed} = \text{frequency} \times \text{wavelength}$$

$$v = f \times \lambda$$ where v = speed in m/s
f = frequency in hertz
λ = wavelength in m

This means that there are two equations for finding the speed of waves. The choice of equation will depend on the information given in the question.

Wave calculations

We can use the wave equations to find some of the features of waves.

Example
Water waves travel from one side of a pond to the other, a distance of 16 m in 4 s. The distance between successive crests is 8 cm. Calculate the frequency of the waves.

Solution
The speed can be found from:

$$v = \frac{d}{t}$$ where $d = 16\,\text{m}$ and $t = 4\,\text{s}$

This gives $\qquad v = 4\,\text{m/s}$

The distance between successive crests gives

$$\lambda = 8\,\text{cm} = 0.08\,\text{m}$$

From the equation
$$v = f \times \lambda$$
$$4 = f \times 0.08$$
$$f = \frac{4}{0.08}$$
$$= 50\,\text{Hz}$$

Section 1.1 Summary

◆ The speed of sound in air is less than the speed of light in air.
◆ Waves are one way of transmitting signals.
◆ The key terms for waves are frequency, wavelength, speed and amplitude.

◆ $v = f \times \lambda = \dfrac{d}{t}$

End of Section Questions

For these questions the speed of sound should be taken as 340 m/s.

1 A note on a musical instrument has a frequency of 512 Hz. If the speed of sound is 340 m/s, calculate the wavelength of this note.

2 The wavelength of a high pitched sound is 0.02 m. Calculate the frequency of this sound.

3 For the wave shown below:

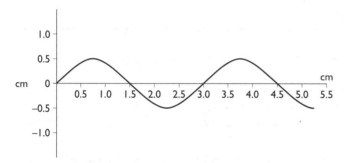

Figure 1.1Q3

(a) What is the amplitude?
(b) What is the wavelength?
(c) If 24 waves pass in 8 s calculate the frequency.
(d) Calculate the speed of the wave.

4 During a thunderstorm the thunder is heard 4 s after the lightning.

(a) Explain why this happens.
(b) How far away is the storm?

At the end of this section you should be able to:

1   Describe a method of sending a message using code (Morse or similar).
2   State that coded messages or signals are sent out by a transmitter and are replayed by a receiver.
3   State that the telephone is an example of long-range communication using wires between transmitter and receiver.
4   State the energy changes (**a**) in a microphone (sound to electrical); and (**b**) in a loudspeaker (electrical to sound).
5   State that the mouthpiece of a telephone (transmitter) contains a microphone and the earpiece (receiver) contains an earphone (loudspeaker).
6   State that electrical signals are transmitted along wires during a telephone communication.
7   State that a telephone signal is transmitted along a wire at a speed greater than the speed of sound (almost 300 000 000 m/s).
8   Describe the effect on the signal pattern displayed in an oscilloscope due to a change in (**a**) loudness in sound; and (**b**) frequency of sound.
9   Describe, with examples, how the following terms relate to sound: frequency and amplitude.
10 State what is meant by an optical fibre.
11 Describe one practical example of telecommunications which uses optical fibres.
12 State that electrical cables and optical fibres are used in some telecommunication systems.
13 State that light can be reflected.
14 Describe the direction of the reflected light from a plane mirror.
15 State that signal transmission along an optical fibre takes place at very high speed.
16 Explain the electrical signal pattern in telephone wires in terms of loudness and frequency changes in the sound signal.
17 Compare some of the properties of electrical cables and optical fibres, e.g. size, cost, weight, signal speed, signal capacity, signal quality, signal reduction per km.
18 State the principle of reversibility of ray paths.
19 Describe the principle of operation of an optical fibre transmission system.
20 Carry out calculations involving the relationship between distance, time and speed in problems on light transmission.

Figure 1.5 Satellite display system on a ship.

Figure 1.6 Enigma coding machine.

Sending signals

Many communication systems often pass information in code. Native North Americans used smoke signals to communicate news of important events. It was a few years later that Samuel Morse developed his system of dots and dashes to send signals using an electrical telegraph system. This Morse code was used for several years. In wartime, flashing lamps were used between ships to avoid detection of electrical activity by the enemy. Ships now use satellites to find and check their position (figure 1.5). During the Second World War the use of the Enigma encoding machine showed how codes could be used with a great deal of mathematical skill (figure 1.6)

The use of codes has not disappeared. Many internet sites use a form of code called encryption, which allows sensitive data, such as financial details, to be encoded when passed to a supplier. This is often denoted by a padlock at the bottom of the internet site. A simple code is the use of text messaging for mobile phones.

The telephone

In 1875, a Scotsman, Alexander Graham Bell sent the first spoken sentence by telephone. He was a teacher for the deaf and had been trying to help people hear messages. When working on his transmitter he spilled some acid on his clothes. He shouted to his friend 'Mr Watson, come here – I want you.' In this way Mr Watson became the first person in the world to hear a message on the telephone. Bell's original machine was a transmitter and receiver. It worked better as a receiver but not as a transmitter. He patented the idea in 1876.

Any communication system needs a transmitter and a receiver and the most common is the telephone. This consists of a microphone which acts as a transmitter (figure 1.7) and a loudspeaker which acts as a receiver. In addition a buzzer or bell indicates when a call is received.

The key parts of the telephone and the energy changes are shown in table 1.2.

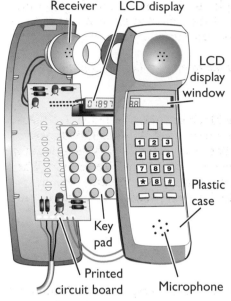

Figure 1.7 Telephone handset.

(labels: Receiver, LCD display, LCD display window, Plastic case, Microphone, Key pad, Printed circuit board)

Part of telephone	Transmitter or receiver	Name of eletrical device inside	Energy changes that take place inside
Earpiece Mouthpiece	receiver transmitter	loudspeaker microphone	electrical to sound sound to electrical

Table 1.2 Parts of a telephone and the energy changes.

During a telephone conversation electrical signals are sent along the communicating wire at almost the speed of light, that is 300 000 000 m/s. These signals can be examined using an oscilloscope. The trace on the oscilloscope shows the electrical patterns of the signals in the wire.

Patterns on the oscilloscope

It is possible to display wave patterns for sound signals on an oscilloscope and to see the effect of changes.

A trace on an oscilloscope screen displays a sound signal as shown in figure 1.8(a). If only the loudness of the sound is increased the pattern on the oscilloscope changes to that shown in figure 1.8(b). If only the frequency of the note is increased the pattern changes to that shown in figure 1.8(c).

These changes show us that:
◆ When the amplitude of a sound changes, the loudness changes but the frequency is unchanged.
◆ When the frequency of a sound changes, the note we hear changes but the loudness is unchanged. The term 'pitch' is used to describe how a noise or musical note sounds to us.
◆ The higher the pitch the higher the frequency.

Sending the signals down the system

Originally any signals that were sent down cables were electrical ones. These signals were sent down thick copper cables. With the growth of communication systems this meant more cables would have to be used. However if signals can be sent as pulses of light then fewer and thinner cables could be used.

The special cables that carry light signals are called optical fibres. To understand how these operate we must look at the reflection of light.

(a)
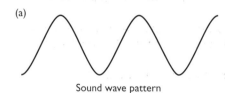
Sound wave pattern

(b)
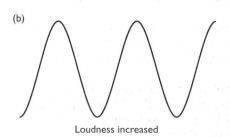
Loudness increased

(c)
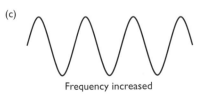
Frequency increased

Figure 1.8 Wave patterns on an oscilloscope.

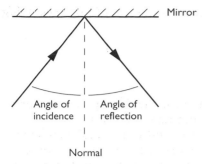

Figure 1.9 Reflection of light.

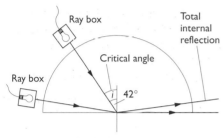

Figure 1.10 Total internal reflection of light.

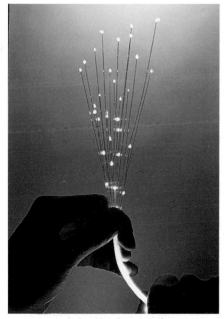

Figure 1.11 An optical fibre is as fine as a human hair.

Law of reflection

A ray of light is shone from a ray box at a plane mirror. The angle of the ray is measured from a line drawn at right angles to the surface called a **normal**. This is the **angle of incidence**. The angle of the reflected ray is measured. This is the **angle of reflection**. This can be repeated for other angles of incidence.

It is found that *the angle of incidence equals the angle of reflection* (figure 1.9).

Total internal reflection

A ray of light is shone into a semi-circular block as shown in figure 1.10. As the angle of incidence is increased then at a certain angle no light will pass from the semi-circular block through the flat face of the block. When no light passes from glass to air we have what is called **total internal reflection**. The smallest angle of incidence at which total internal reflection takes place is called the critical angle. The **critical angle** for glass is about 42°. One use of total internal reflection is in optical fibres.

Optical fibres

Optical fibres are thin flexible glass threads about one-eighth of a millimetre in diameter. Most of the glass thread is an outer layer of glass called cladding glass. In the middle of the thread is a different glass which is less than one hundredth of a millimetre in thickness. The fibres are made of extremely pure glass to cut down light loss. The fibres have a protective surface coating which reflects the light – keeping it inside the fibre.

An optical fibre is formed from glass so pure that a block 36 km thick would be as clear as an ordinary window pane. This means that we could see down to the bottom of the sea! An optical fibre is no thicker than a strand of hair (figure 1.11).

◆ Optical fibres are *lighter, carry more information* (up to one thousand telephone calls per fibre) and give better quality communications than normal telephone wires.
◆ The signal that passes along the fibre is not electrical so it is less likely to be affected by other people's telephone calls or by other forms of electrical interference.
◆ They are cheaper to make than copper since glass is mainly silica which is cheaper than copper.
◆ The disadvantage is that it is more difficult to join fibres together than copper wires.

Many modern telecommunication systems use optical fibres instead of copper wires. One single hair-like fibre can carry all the information needed to bring telephone messages, cable TV and computer services into your home.

Optical fibres operate by light being reflected down the fibre, since no light can leave the outside of the fibre due to the angle of incidence being greater than the critical angle (figure 1.12). It has been estimated that one optical fibre cable could take all the telephone calls being used at once in the world.

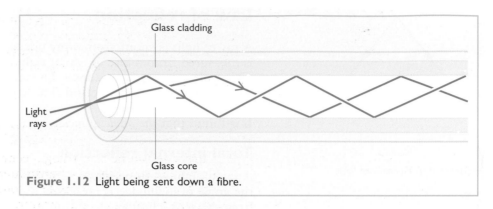

Figure 1.12 Light being sent down a fibre.

Transmission and detection

To send speech information along glass fibres it is first necessary to change sound signals into suitable pulses of electrical energy. The microphone (transmitter) in a telephone handset and some microelectronics do this. These pulses of electricity control a small laser (a narrow, very powerful beam of light; see Chapter 3) or an LED (see Chapter 4) which then produces pulses of light which are transmitted through the optical fibre (figure 1.13).

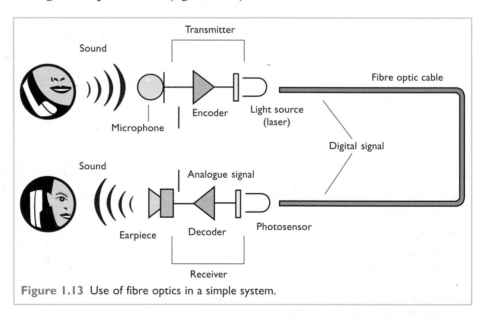

Figure 1.13 Use of fibre optics in a simple system.

At the receiving end the light pulses are changed back into pulses of electricity by a small device called a photodiode. The electrical signal is fed to the earpiece (receiver), which then reproduces the original sound. In 1926, John Logie Baird invented a television system which used optical fibres, but it was not until some 40 years later that Charles Kao and George Hockham suggested that optical fibres might replace copper wires for telecommunications. In 1977 the world's first optical fibre telephone system became operational in America, and in 1978 they were used in a town in Manitoba, Canada to carry telephone, television, radio and computer information.

In 1980 the first trial lengths of submarine optical cables were laid in Loch Fyne and in 1982 an optical fibre link came into operation between London and Birmingham. All of Britain's major cities are now linked by the major fibre link trunk system. Cable TV operators also use these fibres for both TV and telephone communications. A transatlantic optical fibre link is now available. Fibre optic cables are being laid as shown in figure 1.14.

Figure 1.14 Fibre optic cables being laid at sea.

Modern optical fibres transmit light signals with very little signal loss, that is loss of energy, and can be used over distances of about 100 km without amplification. With conventional copper cables there is so much loss (or 'attenuation') of the signal that repeater amplifiers have to be installed every 4 km.

Section 1.2 Summary

- A message can be sent using a code (Morse or similar).
- Signals are sent out by a transmitter and are replayed by a receiver.
- The telephone is an example of long-range communication using wires between transmitter and receiver.
- The energy changes in a telephone are: in a microphone (sound to electrical) and in a loudspeaker (electrical to sound).
- The mouthpiece of a telephone (transmitter) contains a microphone and the earpiece (receiver) contains an earphone (loudspeaker).
- Electrical signals are transmitted along wires during a telephone communication.
- A telephone signal is transmitted along a wire at a speed that is very much greater than the speed of sound (almost 300 000 000 m /s – the speed of light).

- An optical fibre is a thin piece of glass.
- Electrical cables and optical fibres are used in some telecommunication systems.
- Light can be reflected from a mirror and the angle of incidence is equal to the angle of reflection.
- Signal transmission along an optical fibre takes place at a very high speed.

End of Section Questions

1 In parts of Africa the game reserves are very large. Communication could be by telephone.

   (a) Suggest a reason for this method rather than light flares.
   (b) Suggest a disadvantage of this communication system.

2 (a) What are the main parts of a telephone?
   (b) State the energy changes in each part.

3 (a) What is meant by an optical fibre?
   (b) State the advantages of using this fibre compared to conventional cable.
   (c) Draw a diagram to show how light is sent along such a fibre.

At the end of this section you should be able to:

1 State that the main parts of a radio receiver are: aerial, tuner, decoder, amplifier, loudspeaker, electricity supply; and to identify these parts on a block diagram.
2 Describe in a radio receiver the function of the aerial, tuner, decoder, amplifier, loudspeaker and electricity supply.
3 State that the main parts of a television receiver are: aerial, tuner, decoders, amplifiers, tube, loudspeaker, electricity supply; and identify these parts on a block diagram of a television receiver.
4 Describe in a television receiver the function of: aerial, tuner, decoder, amplifier, tube, loudspeaker, electricity supply.
5 Describe how a picture is produced on a TV screen in terms of line build-up.
6 State that mixing red, green and blue lights produces all colours seen on a colour television screen.

7 Describe the general principle of radio transmission in terms of transmitter, carrier wave, amplitude modulation, receiver.
8 Describe the general principle of television transmission in terms of transmitter, carrier wave, modulation, video and audio receivers.
9 Describe how a moving picture is seen on a television screen in terms of:
    (a) Line build-up;
    (b) Image retention;
    (c) Brightness variation.
10 Describe the effect of colour mixing lights (red, green and blue).

Radio and television

All communication systems use a transmitter and receiver. Radio and television are examples of very long-range communication which do not need wires between the transmitter and the receiver. These signals travel as waves and so carry energy. They also travel very quickly – their speed in air is 3×10^8 m/s (300 000 000 m/s).

Amplitude modulation

Sounds are transmitted by waves which have a relatively low frequency. The range of the frequencies that humans can hear is from about 20 Hz to 20 000 Hz. It is not practical to convert these low frequencies into radio waves and transmit them as they are, since they will only travel short distances. Radio waves have to have frequencies of hundreds or thousands of kHz, so the radio wave must be made to carry the lower frequency sound wave information. This is done by altering the wave in some way. One way is to alter the amplitude of the radio wave – this is called **amplitude modulation** (AM).

The sound wave is called the **audio frequency** or a.f. wave. The sound wave is changed into an electrical signal at the transmitter (figure 1.15).

The radio wave is also called the **radio frequency** or r.f. wave (figure 1.16).

The r.f. wave can be made to carry the a.f. wave by altering its size or amplitude. For this reason the r.f. wave is sometimes called a 'carrier' wave.

Amplitude modulation is a way of varying the amplitude of a high frequency radio wave so that it carries a low frequency audio wave. This is shown in figure 1.17.

Figure 1.15 Electrical signals carrying voice information. Audio frequency.

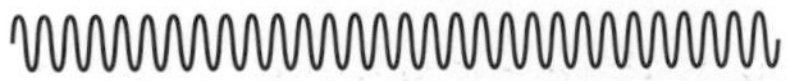

Figure 1.16 Carrier signal. Radio frequency.

Figure 1.17 Amplitude modulated radio signal.

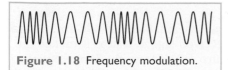

Figure 1.18 Frequency modulation.

In a similar way the frequency of the carrier wave can be altered. This is called **frequency modulation** (FM) and is shown in figure 1.18. The radio stations that use this method can be heard with no interference but the signals do not travel as far as AM signals and more transmitters are needed.

Radio transmission

The parts of a radio transmission system are shown as a block diagram in figure 1.19.

◆ The high frequency radio and the lower frequency audio signals are electrical signals.
◆ The modulator combines the radio signal and the audio signal.
◆ This combined electrical signal is passed on to the amplifier which makes the electrical signal stronger.
◆ The transmitting aerial now changes each electrical signal into a radio wave and sends them out in all directions.

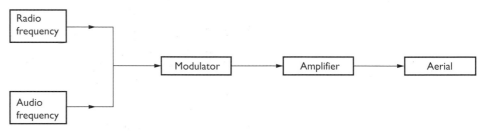

Figure 1.19 Radio transmitter.

Radio reception

The different parts of the radio receiver are shown in figure 1.20.

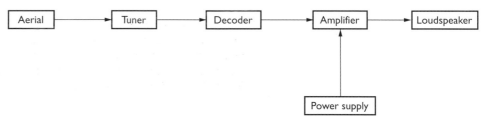

Figure 1.20 A radio receiver.

The aerial

When radio waves arrive at the aerial of a radio receiver they cause the electric charges inside the aerial to move backwards and forwards, in other words to oscillate or vibrate. This movement of electric charges makes a small electric current. The aerial receives *all* of the radio signals and changes them into electrical signals.

Tuner

A radio is able to receive radio waves from many different stations. To keep the signals separate, each station transmits on a different wavelength (and therefore different frequency). The tuner selects the particular radio station you want to receive.

The decoder

When a radio signal is picked up by a receiver it has to be decoded so that it can be changed into speech or music. This separates the higher frequency carrier wave from the lower frequency audio signal. The audio signal contains the information that we want.

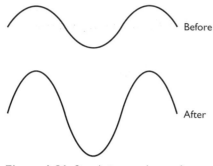

Figure 1.21 Signals into and out of amplifier.

Before

After

The amplifier

The signal received by the radio aerial has a very small amplitude and has to be amplified. The amplifier increases the received signal, that is it makes the amplitude of the received signal larger. Figure 1.21 shows the input signal to the oscilloscope before and after amplification. The amplifier has made the signal bigger and so it now has more energy. This extra energy is supplied from the battery or electrical supply connected to the amplifier. The amplitude (size) of the output signal is controlled by the variable resistor on the amplifier unit. On a radio this is called the 'volume control'.

Listening

The decoded and amplified signal must now be turned into sound energy so you can hear it. This requires a loudspeaker. The loudspeaker changes electrical energy into sound energy.

◆ When the frequency is increased the diaphragm of the loudspeaker vibrates rapidly and a high-pitched note is produced.
◆ When the output control is increased, the current through the coil is increased and the vibrations of the diaphragm are much larger, and a louder sound is produced.

The radio spectrum

Table 1.3 illustrates the radio frequency bands in the UK.

Radio waveband	Frequency range	Wavelength range	Use and example(s)
Low frequency (long wave)	30 kHz–300 kHz	10 km–1 km	Medium to long distance. Radio 4 (200 kHz/1500 m)
Medium frequency (medium wave)	300 kHz–3 MHz	1 km–100 m	Both local and distant sound broadcasts. Ship-shore links. Radio Scotland (810 m); Radio Forth (194 m)
High frequency (short wave)	3 MHz–30 MHz	100 m–10 m	Long distance communication, ship-shore links, navigation, radio beacons, amateur radio C.B. Citizen's Band (27 MHz, FM)
Very high frequency (VHF)	30 MHz–300 MHz	10 m–1 m	Short distance communication, FM sound broadcasts often in stereo. Local radio station Radio Clyde (102.5 MHz)
Ultra high frequency (UHF)	300 MHz–3000 MHz (3GHz)	1 m–10 cm	Air-air, air-ground, short distance communication, colour TV 625 lines BBC 1 and 2, ITV and Channel 4
Super high frequency (SHF)	3 GHz–30 GHz	10 cm–1 cm	Microwave, point-point communication. Radar, satelites
1 GHz = 1 gigahertz = 1 000 000 000 Hz, 1 MHz = 1 megahertz = 1 000 000 Hz			

Table 1.3 Radio frequency bands in the UK.

Example
Using information from the table 1.3 what is the wavelength of Radio Scotland? Calculate the frequency and in which waveband is it?

Solution
From the table, wavelength = 810 m. To calculate the frequency we need to use the equation:

$$v = f \times \lambda \quad \text{with } v = 3 \times 10^8 \, \text{m/s}$$
$$3 \times 10^8 = f \times 810$$
$$f = \frac{3 \times 10^8}{810}$$
$$= 3 \times 10^5 \, \text{Hz}$$
$$= 300 \, \text{kHz}$$

From the table, this frequency is in the medium waveband.

Digital radio

Fascinating Physics

Digital radio is the latest development. It is sometimes called DABS (Digital Audio Broadcasting System). The receiver is expensive but has certain advantages:
- In FM reception, there is interference caused by the signals reflecting off buildings and hills. Using a special frequency-splitting technology this is eliminated and the same frequency can be used across the country. This means that you do not need to retune your set as you move to another area.
- The sound quality is noticeably better.
- There will be a wider range of stations. You will be able to receive 16 national and 16 local stations.

Television transmission

The parts of a television (TV) transmission system are shown in the block diagram (figure 1.22).

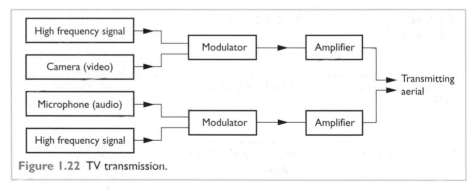

Figure 1.22 TV transmission.

- The high frequency, video and audio signals are electrical signals.
- The modulator combines the high frequency signal, the video signal and the audio signal.
- This combined electrical signal is sent to the amplifier which makes the electrical signal stronger.
- The aerial changes this electrical signal into radio and TV waves which are transmitted in all directions.

The television receiver

The main parts of a TV are detailed below and shown in figure 1.23.
- The aerial picks up wave energy with many different frequencies.
- The tuner picks out the frequency of the TV station you want.

The audio and visual signals are sent out separately by the transmitter. This sometimes means that although we can see a programme we cannot hear the sound due to a fault in the transmission.

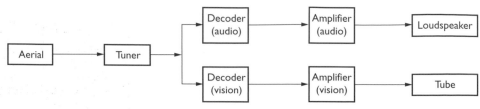

Figure 1.23 Reception.

Batteries or mains electricity is required to make the television receiver operate.

◆ The energy changes in the loudspeaker are electrical to sound.
◆ The energy change that takes place at the TV screen is electrical to light.

The picture tube

The electron gun, which is at the narrow end of the tube, produces negative charges (electrons) and 'fires' them as an invisible beam. As they cannot travel very far in air, there is a vacuum (no air) inside the tube to allow them to reach the screen. A fluorescent coating on the screen produces a tiny spot of light when hit by the electrons (figure 1.24).

This spot can be moved around the screen by deflecting the electron beam. In a TV picture tube, electromagnets are used. These are magnets powered by electricity and are coils of wire carrying a current (see Chapter 2). One pair of coils is arranged to move the beam up and down (vertically) and another pair arranged to move it from side to side (nearly horizontally).

Because the electrons in these tubes comes from a part of the electron gun called the cathode, the tubes are sometimes called **cathode ray tubes**.

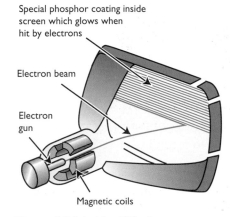

Special phosphor coating inside screen which glows when hit by electrons

Electron beam

Electron gun

Magnetic coils

Figure 1.24 Inside a TV tube.

How a television picture is formed

A British television picture is built up from 625 lines but in America it is 525 lines. The more lines the better the definition, or clarity, of the picture. The electron beam begins at the top left corner and moves across the screen at a speed of about 7000 m/s.

On completing the line the beam is switched off, and moves to the left side, before starting the next one. The lines which build up the picture are not built up in sequence but are done as lines 1, 3, 5, etc. followed by the other lines 2, 4, 6, etc. This technique is called 'interlacing'.

This continues until all 625 lines are completed so forming one picture. The beam then returns to the top left corner to start on the next picture (figure 1.25).

The TV signal from the aerial controls the brightness of the spot, by altering the number of electrons which travel from the electron gun to the screen. The greater the number of electrons, the brighter the spot.

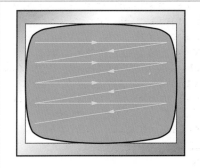

Figure 1.25 Time taken for the beam to travel from the top to the bottom of the screen is 0.025 seconds.

Moving pictures

The eye takes about 1/10th of a second (0.1 s) to become aware of an object suddenly placed in front of it, and this vision persists for about 1/10th of a second after the object has disappeared. When the brain receives pictures quickly one after another it holds each picture for a short time – known as 'image retention or persistence of vision.' If the next picture appears before the previous one disappears the brain puts the pictures together and is not aware of a space in between. This is what happens at the cinema or on a TV screen because the pictures follow each other so quickly.

In TV there are 25 pictures every second (one picture every 0.04 s) and, in the cinema, 24, each one slightly different from the one before. The effect of the persistence of vision is of continuous movement rather than a set of still pictures (figure 1.26).

Figure 1.26 A series of still pictures from a film, showing persistence of vision.

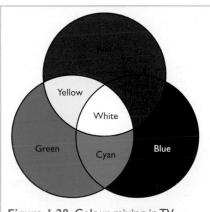

Figure 1.27 Shadow mask (colour television).

Figure 1.28 Colour mixing in TV pictures.

The colour TV tube

A colour TV has three electron guns and a screen coated with about one million tiny dots arranged in triangles. When these dots are hit by electrons, one dot in each triangle gives out red light, another green light and the third blue light. It is the material of the dots which give out colour. *There are no coloured electrons.*

As the three electron beams sweep or scan the screen, an accurately placed 'shadow mask' makes sure that each beam strikes only dots of one colour, for example electrons from the 'red' gun only hit 'red' dots. The shadow mask consists of a metal plate with about one third of a million holes on it. When a triangle is struck it may be that the red and green electron beams are very strong (intense) but not the blue. The triangle will give out red and green light strongly and appear yellowish. The triangles are struck in turn and since the dots are so small and the sweeping so fast, we see a continuous colour picture (figure 1.27).

The mixing of the three different colours can produce other colours.

◆ Red light and blue light give magenta light.
◆ Blue light and green light give cyan light.
◆ Green light and red light give yellow light.
◆ Red, blue and green lights give white light (figure 1.28).

Three electron guns fire separate invisible beams of electrons and magnetic coils move them up or down, while other coils move the beams sideways. These are moved simultaneously and the electrons can be aimed at any point on the screen. The screen is covered in dots, which are made of three special materials which glow when they are hit by electrons. A colour television screen is made up of 625 lines, each line having around 800 sets of dots. Different colours are made by changing the brightness of the red, blue and green dots. With each separate picture, the colour pattern may change. This creates the effect of a moving picture due to image retention.

Televisions with no tubes

Most people have seen the new flat square tubes which are often used with modern televisions. The screen is not curved as in other sets and the magnets inside deflect the electrons to give a more natural view. They also use a rectangular view similar to that of the cinema screen.

Fascinating Physics

Figure 1.29 Plasma TV set.

Many computers now use a flat display rather like a picture frame, which uses a liquid crystal display (LCD) rather than a tube. Some screens are available as LCD screens similar to the screens on laptop computers. However they have better viewing angles than those of laptop computers. The latest ones are called 'plasma' tubes. Plasma tubes have the advantage of reducing the space required but at present are about ten times more expensive and the sets are very heavy. These sets are shown in figure 1.29.

Section 1.3 Summary

◆ The main parts of a radio receiver are: aerial, tuner, decoder, amplifier, loudspeaker, electricity supply.

◆ The main parts of a television receiver are: aerial, tuner, decoders, amplifiers, tube, loudspeaker, electricity supply.

◆ A picture is produced on a TV screen in terms of line build-up with 625 lines needed.

◆ Mixing red, green and blue lights produces all colours seen on a colour television screen.

◆ Amplitude modulation occurs when a carrier wave is combined with an audio or video signal. This allows the combined wave to be sent considerable distances.

◆ A moving picture is seen on a television screen in terms of: line build-up, image retention and brightness variation.

End of Section Questions

1 A student writes that a radio only needs an aerial, a tuner and a small amplifier and power supply to operate.

(a) Which two parts are missing?
(b) Explain the purpose of one of the missing parts.

2 A radio operates at 92.5 MHz in the FM band.

(a) What is meant by the term FM?
(b) Calculate the wavelength of this station.
(c) Use table 1.3 above to state the band that it operates in.

3 (a) State the three main colours that make up a TV screen.
(b) How would a colour like purple be made from these basic colours?

4 The latest TV sets can have over 1200 lines to make up the picture. Explain how the line build-up makes a complete picture on the screen.

5 Some of the latest videophones used in news broadcasts make the picture appear to be moving in a series of jumps. This is because the picture is taken at 15 frames per second rather than 25. Explain why the number of frames per second gives an illusion of a moving picture.

6 Amplitude modulation is needed to send signals from any radio station.

(a) What is the advantage of this technique?
(b) In the radio receiver the reverse process of demodulation occurs. What happens in this part of the radio?

At the end of this section you should be able to:

1 State that mobile telephones, radio and television are examples of long-range communication which do not need wires (between transmitter and receiver).
2 State that microwave television and radio signals are waves which transfer energy.
3 State that microwaves, television and radio signals are transmitted at very high speed.
4 State that microwave, television and radio signals are transmitted through air at 300 000 000 m/s.
5 State that a radio transmitter can be identified by wavelength or frequency values.
6 State that curved reflectors on certain aerials or receivers make the received signal stronger.
7 Explain why curved reflectors on certain aerials or receivers make the received signal stronger.
8 Describe an application of curved reflectors used in telecommunications, e.g. satellite TV, TV link, boosters, repeaters or satellite communication.
9 State that the period of satellite orbit depends on its height above the Earth.
10 State that a geostationary satellite stays above the same point on the Earth's surface.
11 Describe the principle of transmission and reception of satellite television broadcasting using geostationary satellites and aerials.
12 Describe the principle of intercontinental telecommunications using a geostationary satellite and ground station.

13 Carry out calculations involving the relationship between distance, time and speed in problems on microwaves, television and radio waves.
14 Carry out calculations involving the relationship between speed, wavelength and frequency for microwaves, television and radio waves.
15 Explain some of the differences in properties of radio bands in terms of source strength, reflection, etc.
16 Explain in terms of diffraction how wavelength affects radio and television reception.
17 Explain the action of curved reflectors on certain transmitters.

Diffraction of waves

Diffraction is a bending of waves and happens when waves go through a very narrow gap or round an obstacle. You can observe this effect as water waves enter a harbour opening. A typical effect is shown in figure 1.30.

The amount of diffraction of a wave depends on the wavelength. The larger the wavelength (the smaller the frequency), the more the diffraction. This also applies to radio and TV waves, that is long wavelengths (low frequencies) are diffracted more than short wavelengths (high frequencies).

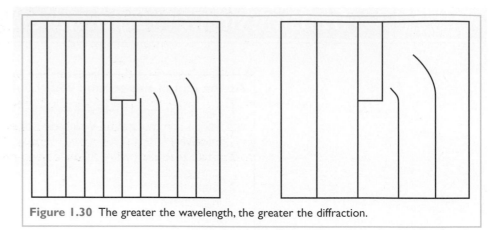

Figure 1.30 The greater the wavelength, the greater the diffraction.

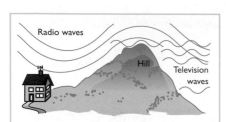

Figure 1.31 Long and medium wavelength radio waves bend around hills much better than short wavelength TV waves. Therefore, this house will have poor TV reception.

Diffraction of radio and TV waves

TV waves, having a shorter wavelength than radio waves, are diffracted less, therefore are more difficult to pick up in certain areas such as a glen (figure 1.31). This is because there is lower strength of signal received by the television receiver since there is less diffraction.

Mobile phones

There are about 30 million mobile phones in use in the UK. They range from simple communication devices, to internet WAP phones (Wireless Application Protocol). There are also phones which allow much faster connection for internet sites than the slower connections at present. Each phone is a miniature radio transmitter and receiver which operates by batteries (figure 1.32). The key features are contained within the SIM card which has your unique number and the calling plan details from your phone company. The idea is not new but the problem faced was to have a large number of users within a limited amount of frequencies. To avoid interference, each call must be made on its own frequency, but each user could not be given their own frequency since this would use up large amounts of the frequency range.

The solution is to divide the country into a number of areas called cells. Within each cell, one precise group of frequencies called channels is used to both transmit and receive calls (figure 1.33). The same set cannot be used in a neighbouring cell since calls would interfere with each other, but they can be used in a cell that is further away (figure 1.34).

Figure 1.32 Circuit board in mobile phone.

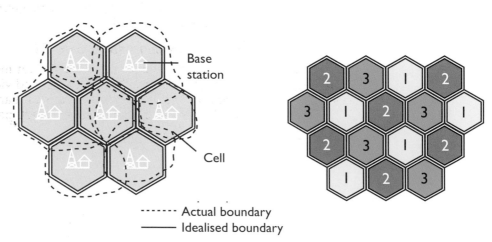

Figure 1.33 Different cells for mobile phones.

Figure 1.34 Allocation of frequencies.

The frequencies used are UHF, that is above 300 MHz. Two of the providers operate on 900 MHz and the other two on 1800 MHz. These frequencies have suitable transmission characteristics:
◆ A short useful range.
◆ Work best when there is clear line of sight between the transmitter and receiver.
◆ Have waves that can be directed very precisely and need only small aerials.

To help with the growth of these phones, the cells in a city area are much smaller than those in the country areas. The phones change frequencies as you move from one edge of the cell to another as shown in figure 1.35.

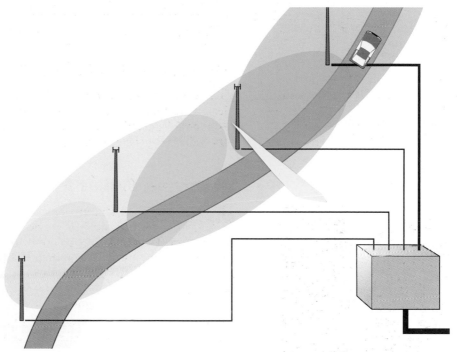

Figure 1.35 Signal strength.

There are some disadvantages of mobile phones:
◆ They cannot pick up signals in some hilly areas.
◆ They cannot receive signals inside metal places like a lift.
◆ There may be a danger since there is evidence that, since they generate heat, the microwaves used can cause some brain damage if they are used for long periods of time. There is information available on the radiation emitted by different phones. This gives the radiation absorbed per kilogram of the body.

Internet ready and video phones G3

Fascinating Physics

The latest telephones can transmit not just e-mails but information from a wide range of sites. They can also send video links – but they are difficult to use and have probably more features than most people will ever need. The manuals for these phones are over 900 pages (figure 1.36).

Figure 1.36 This G3 mobile phone transmits video.

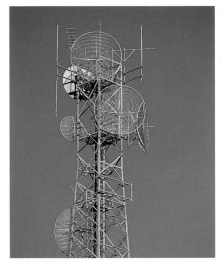

Figure 1.37 Transmitting dish aerial.

Receiving signals from satellites

Dish aerials

If a transmitting aerial is placed at the focus of a curved reflector (or dish), the reflected signal from the dish has the shape of a narrow beam (just like the light from a torch). This allows a strong (concentrated) signal to be sent in a particular direction from the transmitting dish aerial. (figure 1.37).

Using dish aerials (curved reflectors)

When radio broadcasting began in 1920 the medium frequency (MF) radio signals then used could travel about 1600 km. Soon after, high frequency (HF) radio bands were discovered and these were used for world-wide communication.

Very high frequencies (VHF) soon followed, but these were unable to travel round the curvature of the Earth (the higher the frequency, the smaller the wavelength and the smaller the diffraction). Today VHF is used mainly for mobile communication, for example, between aircraft and ground stations. As more and more information was transmitted the frequencies mentioned above became overcrowded and even higher frequencies had to be found. This led to the use of microwaves whose frequency is about 10^9 Hz (1 GHz). Microwaves were useful since large amounts of information could be sent but little power was required, and it was also found that they were easily focused using dish aerials (figure 1.38(a)).

Receiving dishes gather in most of the signal and reflect it to one point called the focus. The receiving aerial is placed at the focus to receive the strongest signal (figure 1.38 (b)).

Microwaves are unable to diffract (bend) round obstacles because of their small wavelength. This means that the transmitting and receiving dish aerials must be in line of sight and are often located on towers. Since microwaves have a fairly short range, a series of repeater (relay) stations are needed every 40 km. The incoming microwaves are focused onto the receiving dish aerial. Then after being amplified (made bigger) they are passed to the aerial at the focus of the transmitting dish to produce a parallel narrow beam, which is sent to the next repeater station. Using

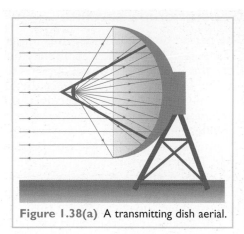

Figure 1.38(a) A transmitting dish aerial.

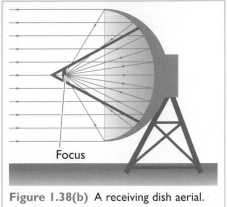

Focus

Figure 1.38(b) A receiving dish aerial.

microwaves for round the world communication would require several hundred repeater (relay) stations at ground level. However only three satellite repeater stations in the sky are required to cover all the earth if they are in the correct positions. This was predicted by Arthur C. Clarke in 1945.

Satellites

Figure 1.39 A satellite receiving dish at Goonhillie Down, Cornwall.

A satellite has a very sensitive receiver as microwaves have to travel about 36 000 km to reach it. The signal when received and focused, is amplified and transmitted back to Earth. Very large dish aerials on the Earth are required to pick up the weak signals coming from satellites and also to send signals accurately to them (figure 1.39). The dish to receive the signals is the same design as the transmission one but the signals are received in a parallel beam and the rays are then brought to a focus. Figures 1.38(a) and (b) show that the diagram is the same, but the direction of the arrows is reversed. A typical satellite is shown in figure 1.40.

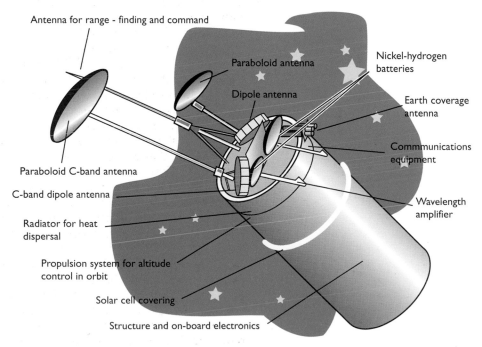

Figure 1.40 Intelsat satellite.

Geostationary satellite

The period of a satellite depends on its height above the Earth's surface. (The period is the time to complete one orbit.) The further the satellite is from the Earth, the slower it appears to move. It has been calculated that a satellite in orbit 36 000 km above the equator would complete one orbit in the same time as the Earth revolves, that is 24 hours. This is called a geostationary satellite as it appears to be stationary above the Earth. Satellites in geostationary orbit must be in orbit above the equator. Some satellites are placed in higher orbits.

In 1965 Early Bird was placed in a geostationary orbit. With its receiving dish aerial it picked up microwave signals, focused, amplified and transmitted them as a narrow parallel beam back to Earth. Panels of solar cells provided the energy to do this.

The satellite communication system has expanded rapidly. In 1971 there were 13 satellites in geostationary orbit and over 200 Earth satellites. Each new satellite is an improvement on the previous ones. Early Bird could relay 240 telephone calls at once. Modern satellites have several thousand telephone and data channels and television channels.

Sending a signal

When you pick up the telephone and use it to call America there is a complex sequence of events that take place. This is shown in figure 1.41. It is important to note that the satellite does not act as a mirror but is a combined receiver and transmitter.

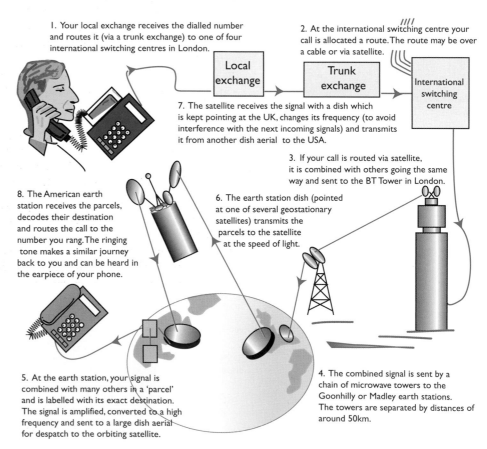

1. Your local exchange receives the dialled number and routes it (via a trunk exchange) to one of four international switching centres in London.

2. At the international switching centre your call is allocated a route. The route may be over a cable or via satellite.

7. The satellite receives the signal with a dish which is kept pointing at the UK, changes its frequency (to avoid interference with the next incoming signals) and transmits it from another dish aerial to the USA.

3. If your call is routed via satellite, it is combined with others going the same way and sent to the BT Tower in London.

8. The American earth station receives the parcels, decodes their destination and routes the call to the number you rang. The ringing tone makes a similar journey back to you and can be heard in the earpiece of your phone.

6. The earth station dish (pointed at one of several geostationary satellites) transmits the parcels to the satellite at the speed of light.

5. At the earth station, your signal is combined with many others in a 'parcel' and is labelled with its exact destination. The signal is amplified, converted to a high frequency and sent to a large dish aerial for despatch to the orbiting satellite.

4. The combined signal is sent by a chain of microwave towers to the Goonhilly or Madley earth stations. The towers are separated by distances of around 50km.

Figure 1.41 You can see what happens in the time between dialling the first digit and hearing the ringing tone in New York.

Browsing the web

Perhaps you have used the internet to find a web site like NASA which will give you the latest on the shuttle. The internet is a computer linked system which connects through the telephone system to other computers world-wide. To do this you need a MODEM (a MODulator DEModulator). This is a device which changes the computer signals into an electrical form which can be sent down a phone link and can do the reverse task for the incoming signals. The different parts of the information system are called web sites.

Find out about a few web sites that you can visit.

Figure 1.42 A cable television control room.

Cable television

With the growth in television channels some companies are now offering a cable to bring these channels into your home. This means that there is no need for a satellite dish and you can still receive more channels. The cable company has to dig up the roads to lay the cables but you can get the satellite TV channels. The cable company uses large satellite dishes to receive the programmes and then distributes them down a cable system using fibre optics to keep the signal as strong as possible. The advantages are that with more channels there is the extra capacity in the system which would be difficult with a small dish (figure 1.42).

Section 1.4 Summary

- Mobile telephones, radio and television are ⬦⬦⬦les of long-range communication which do not need wires (bet⬦⬦ ⬦nsmitter and receiver).
- Microwave, television and radio signals are waves which ⬦sfer energy and are transmitted at a very high speed namely 300 000 ⬦
- A radio transmitter can be identified by wavelength or frequency values.
- Curved reflectors on certain aerials or receivers make the received signal stronger by bringing the rays to a focus.
- Some uses of curved reflectors used in telecommunication are satellite TV, TV link, boosters, repeaters or satellite communication.

- The period of satellite orbit depends on its height above the Earth.
- A geostationary satellite stays above the same point on the Earth's surface.
- Diffraction is a bending of the waves around a gap or obstacle. The amount of diffraction increases with wavelength.

1 (a) State one advantage and one disadvantage of using mobile phones.
  (b) When you are between two buildings the mobile phone signal is faint. Suggest a reason for this.

2 All satellite TV companies require curved dishes, rather than flat aerials, to receive signals. Explain with a diagram why this is necessary.

3 A signal is sent via a satellite. It is sent up at one frequency and then retransmitted at a different frequency. Explain why two different frequencies are used.

4 Weather satellites have a polar orbit which is close to the Earth but communication satellites have orbits around the equator. What does this tell you about the height and orbit time of a communications satellite compared to a weather one?

5 TV signals have a higher frequency than radio signals. Explain why TV signals are not picked up beyond a hill but some radio signals are received.

1 A newspaper has the information for radio stations as follows:

Radio 4
FM 92.4 – 94.6 MHz
LW 198 kHz (1515 m)

  (a) The 'F' in FM means frequency. What does the 'M' in the term represent ?
  (b) The frequency is 198 KHz on long wave. What does the term 'frequency' mean?
  (c) Show by calculation that the wavelength on LW is 1515 m.

2 Water waves travel from one end of a pool to another. The length of the pool is 48 m.

  The waves take 6 s to reach the end of the pool.

  (a) Calculate the speed of the waves.
  (b) If the frequency of the waves is 6 Hz, calculate the wavelength of the waves.

3 Colour TVs contain three electron guns which produce a colour picture on the screen

  The guns are contained within a vacuum tube

  (a) State the three different colours of the guns
  (b) Explain how a single frame is produced on the screen
  (c) How is a moving picture produced?

4 In a radio there are several parts. Explain the function of the following parts:

  (a) The tuner
  (b) The amplifier
  (c) The demodulator is used since the signal received by the radio is a combination of two different signals. Explain what the demodulator does and draw diagrams to illustrate the different signals

5 Mobile phones operate on different frequencies depending on the company. The two main frequencies are 900 MHz or 1800 MHz.

  (a) Calculate the wavelength of the 1800 MHz signal
  (b) As you pass into a valley between two hills the signals fade. This is due to the bending of the waves. State what this effect is called
  (c) Why will the 900 MHz operating frequency show the largest amount of bending?

6 Satellites orbit the earth at different heights. Polar satellites take a shorter time to orbit compared to communications satellites

  (a) How do the heights of these two satellites compare?
  (b) Communications satellites are geostationary. Explain what is is meant by a geostationary satellite
  (c) Such a satellite orbits at a height of 36 000 km. Calculate the time taken for a signal to go up and then return to Earth.

7 Fibre optic systems are used in telecommunication systems

  (a) What is meant by an optical fibre?
  (b) A length of fibre optic cable is 300 km long. If the speed of light in the cable is 2×10^8 m/s calculate the time for the signal to reach the receiver.

CHAPTER TWO
Using Electricity

> **At the end of this section you should be able to:**
>
> 1 Describe the mains supply/battery as a supply of electrical energy and describe the main energy transformations occurring in household appliances.
> 2 State approximate power ratings of different household appliances.
> 3 Select an appropriate flex given the power rating of an appliance.
> 4 State that fuses in plugs are intended to protect flexes.
> 5 Select an appropriate fuse given the power rating of an appliance.
> 6 Identify the live, neutral and earth wire from the colour of their insulation.
> 7 State to which pin each wire must be connected for plug, lampholder and extension socket.
> 8 State that the human body is a conductor of electricity and that moisture increases its ability to conduct.
> 9 State that the earth wire is a safety device.
> 10 State that electrical appliances which have the double insulation symbol do not require an earth wire.
> 11 Draw the double insulation symbol.
> 12 Explain why situations involving electricity could result in accidents (to include proximity of water, wrong fuses, wrong, frayed or badly connected flexes, short circuits and misuse of multiway adaptors).
> 13 Explain how the earth wire acts as a safety device.
> 14 Explain why fuses and switches must be in the live wire.

Introducing electricity

It is not very often that we have a 'blackout' but when it does happen, as a result of the mains electricity failing, what a disruption and inconvenience it causes – yet the convenience of having mains electricity in our homes only became normal to the majority of people in this country during the 1920s and 1930s.

We are now used to so many electrical appliances at home for cooking, cleaning, heating and entertainment, it's hard to imagine life without them. However, before the pressing of a switch can activate an appliance, the electricity has to be produced and transmitted to our homes – this is explained in Chapter 6.

An understanding of what electricity is and how it can be safely used are important elements of this chapter. Always remember:

mains electricity can kill!

The electrical appliances we use in our homes, industry and schools use **electrical energy** from the mains supply or from batteries. They change electrical energy into a form which is useful for what we are doing (and in

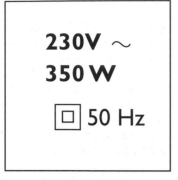

food mixer rating plate

230 V ∿

50 Hz

2200 W

Made in Gt.Britain

kettle rating plate

230 V ∿ 950W

50 Hz

Made in Gt.Britain

toaster rating plate

Figure 2.1 Typical rating plates for a food mixer, kettle and toaster.

Figure 2.2 Double insulation symbol.

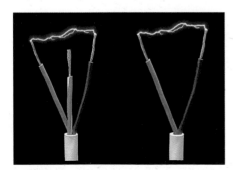

Figure 2.3 Three-core and two-core flexes.

many cases into other forms of energy which are not so useful). For instance, a lamp changes electrical energy into heat energy and light energy. For a lamp, the main energy change, and the most useful is the change of electrical energy into light energy.

The list below shows the main energy change for a number of household appliances:

◆ A kettle changes electrical energy into **heat energy**.
◆ A radio changes electrical energy into **sound energy**.
◆ A lamp changes electrical energy into **light energy**.
◆ A washing machine changes electrical energy into **kinetic (movement) energy**.

Flexes, plugs and fuses

Let's look at three common household appliances – a food mixer, a kettle and a toaster. They are fitted, like most electrical appliances, with **rating plates** and these are shown in figure 2.1. The rating plate gives information about the appliance, that is the voltage (the number with the letter 'V' after it) and frequency (the number with the letters 'Hz' after it) required to operate it, its power rating (the number with the letter 'W' after it) and the symbol shown in figure 2.2 if it is double insulated.

The electric kettle has the highest power rating at 2200 **watts** (**W**) and because of this the flexible cord, or flex, which connects it to the mains wall socket will, generally, be the thickest. A flex consists of two or three cores of thin, stranded, insulated copper wire within an outer insulating sheath as shown in figure 2.3. The brown covered core is the **live** wire, the blue covered core is the **neutral** wire and, in three-core flexes, the green and yellow striped covered core is the **earth** wire.

The cores in a flex, during normal use, heat up when a current passes through them. However, if they carry too high a current they will become too hot and overheat, and this could cause a fire. To protect the flex from too high a current, the three-pin plug connected to the flex is fitted with a **fuse**. A fuse is simply a thin piece of wire. When an electric current passes through it, it heats up. If the current gets too high, then the fuse wire will become so hot that it melts or 'blows', breaking the electrical circuit and so protecting the flex.

Fuses for three-pin plugs, known as **cartridge fuses**, are available in a number of sizes – the most common being 3 A (ampere) and 13 A.

There are a number of different types of flex available for different applications. They differ in the maximum safe current they can carry (which depends on the thickness of the conductors making up the core), the number of cores, the type of outer insulation and in their cost. Table 2.1 shows the minimum thickness of conductor required for appliances of different power ratings.

Power rating	Typical appliance	Thickness of conductor	Maximum current
Up to 700 W	clock, food mixer	0.50 mm^2	3 A
700 to 1380 W	hairdryer, toaster	0.75 mm^2	6 A
1380 to 2300 W	kettle, fan heater	1.00 mm^2	10 A
2300 to 3000 W	3 kW heater	1.25 mm^2	13 A

Table 2.1 Minimum thickness of conductor for different appliances.

Choosing a fuse

To select the correct size of fuse the power rating or wattage (W) of the appliance, marked on the rating plate, must be known. Generally, if the power rating of the appliance is less than 700 W then a 3 A fuse should be fitted. For a power rating greater than or equal to 700 W a 13 A fuse should be fitted. (Some appliances such as refrigerators and freezers need a larger fuse than the wattage on the rating plate indicates. These appliances have an electric motor. When the electric motor switches on, an initially high current passes which would 'blow' a 3 A fuse. In these cases the instructions given by the manufacturer with regard to the fuse value to be fitted in the three-pin plug *must* be followed.)

Table 2.2 shows the power ratings and cartridge fuse values for the three appliances given in figure 2.1.

Appliance	Power rating	Fuse
food mixer	350 W	3 A
kettle	2200 W	13 A
toaster	950 W	13 A

Table 2.2 Power rating and fuses for electrical appliances.

Wiring a three-pin plug

For safety, it is very important that a three-pin plug is correctly wired (figure 2.4). Provided the appliance is working properly, then electrical current only passes through the live and neutral cores of the flex – no current passes through the earth wire. The earth wire is connected to the outer metal casing of the appliance and to the earth pin in the three-pin plug. Its purpose is to provide a very easy path for electrical current to pass to earth should a fault occur and it may be considered as a 'safety' wire.

However, if the appliance develops an electrical fault, such as the one shown in figures 2.5(a) and (b) then the earth wire is very important. In (a) a large current will pass through the live and earth wires. This current will be much larger than the fuse value and so the fuse will 'blow', breaking the electrical circuit. The toaster is now safe for anyone touching it as the broken fuse has disconnected it from the live wire.

Neutral wire
Blue

E
L
13 A
fuse

13 A

N

Live wire
Brown

Cable grip

Earth wire
Green/yellow

Flex to TV

Figure 2.4 A correctly wired three-pin plug.

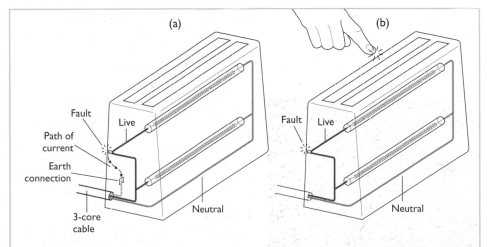

(a)

(b)

Fault
Live

Path of current

Earth connection

Neutral

3-core cable

Fault
Live

Neutral

Figure 2.5 (a) and (b) show a faulty toaster. In (a) the fuse 'blows' when the fault occurs and the toaster can be touched safely. In (b) the earth wire is disconnected and the person receives a shock.

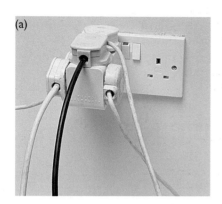

In (b) the earth wire is not connected and so the metal parts of the toaster are 'live'. Anyone touching the toaster will get a shock or possibly be killed – current will pass from the live wire through the person to earth. The danger is greatly increased if the person's fingers are wet as water helps to conduct the electricity through your body.

The plug fuse is fitted to protect the flex from overheating and it must be connected to the live wire. If the fuse 'blows' as a result of the live wire touching a metal part of the appliance then no part of the appliance or flex can remain live. Also, if the appliance has a switch, it must be connected to the live wire so that when it is 'off', no part of the appliance can remain live. Connecting the live and neutral wires the wrong way round would mean that the appliance remains live even when the switch on the appliance is off!

Double insulation

Some appliances, such as hairdryers and electric drills, do not require an earth wire since there are two layers of insulation around their electrical parts. This makes the earth wire unnecessary and so a two-core flex is fitted. The **double insulation** symbol was shown in figure 2.2. Appliances which are double insulated have this symbol shown on their rating plate. Of the three rating plates shown in figure 2.1, the food mixer is the only appliance that is double insulated.

Dangerous situations

◆ The wall socket shown in figure 2.6(a) has too many appliances plugged into it. This could result in too much current being drawn from the wall socket, which could lead to the socket overheating.
◆ When your body is damp or wet, its ability to conduct electricity is greatly increased. Touching wall sockets or light switches with wet or damp hands can be hazardous as the water makes you a better conductor. If water entered the socket or light switch this could provide a path for current to pass through your body and you could be electrocuted. It is for this reason that bathrooms do not have wall sockets and the light switch is either fitted with a cord inside the bathroom or placed on a wall outside.
◆ Frayed, worn or joined flexes are all dangerous (figure 2.6(b)), since the live core may become bare and someone may touch it. In these cases a replacement flex of the correct rating should be fitted.

Figure 2.6 (a) With too many appliances connected to it, the socket could overheat and (b) This frayed flex makes the wire dangerous.

Section 2.1 Summary

◆ Appliances with a power rating up to 700 W are normally fitted with 3 A fuses.
◆ Appliances with a power rating greater than 700 W are normally fitted with 13 A fuses.
◆ Fuses are intended to protect the flex of an appliance from overheating.
◆ Three-core flexes consist of live (brown), neutral (blue) and earth (green and yellow) wires.
◆ Two-core flexes, used with double insulated appliances, consist of live (brown) and neutral (blue) wires.
◆ The earth wire is a safety device.
◆ Fuses and switches are connected in the live wire.

1 Write down the main energy change for each of the following household appliances:

   (a) Electric fire;
   (b) Electric food mixer;
   (c) Radio;
   (d) Electric iron.

2 The diagram below shows four rating plates.

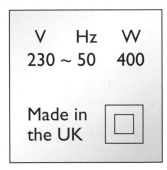

electric drill rating plate

hi-fi rating plate

Serial No
974356

230 V ~
50 Hz

refrigerator rating plate

2200 W
to 2400 W

~ 220 to 240 V
50 Hz

kettle rating plate

Figure 2.1Q2

   (a) Which appliance(s) are double insulated?
   (b) Which appliance(s) require a three-core flex?

3 What value of fuse, 3 A and 13 A, should be fitted in the three-pin plug of a:

   (a) 60 W table lamp;
   (b) 1400 W vacuum cleaner;
   (c) 300 W computer system;
   (d) 1100 W toaster?

4 Using table 2.1 on page 28, select the minimum safe thickness of conductor in a flex that would be required for:

   (a) an electric drill with a power rating of 460 W;
   (b) a kettle with a power rating of 2200 W.

5 Explain why it would be dangerous if bathrooms were fitted with three-pin sockets.

6 A microwave oven develops an electrical fault. The metal outer casing of the microwave becomes live as a result of this fault. Explain how the earth wire acts as a safety device when this fault occurs.

At the end of this section you should be able to:

1 State that the mains supply is a.c. and a battery is d.c.
2 Explain in terms of current the terms a.c. and d.c.
3 State that the frequency of the mains supply is 50 Hz.
4 State that the declared value of the mains voltage is quoted as 230 V.
5 Draw and identify the circuit symbol for a battery, fuse, lamp, switch, resistor, capacitor, diode and variable resistor.
6 State that electrons are free to move in a conductor.
7 Describe the electric current in terms of the movement of charges around a circuit.
8 Use correctly the units 'ampere' and 'volt'.
9 State that the declared value of an alternating voltage is less than its peak value.
10 Carry out calculations involving the relationship between charge, current and time.
11 Use correctly the unit 'coulomb'.
12 State that voltage of a supply is a measure of the energy given to the charges in a circuit.

What is electricity?

All solids, liquids and gases, are made up of **atoms**. An atom consists of a positively charged centre or **nucleus** surrounded by a 'cloud' of rapidly revolving negative charges called **electrons**. The nucleus is made of particles called **protons** (positively charged) and **neutrons** (uncharged) – see figure 2.7.

Charge is measured in **coulombs** (C). The charge on a proton is equal in size to that of an electron. This value is 1.6×10^{-19} C.

Figure 2.7 A model of an atom.

Charge, current and time

Consider a simple electrical circuit – a lamp connected to a battery. The lamp lights because negative charges (electrons) from the negative terminal of the battery move through the wires and lamp to the positive terminal of the battery. This movement of negative charges is called an **electric current** (or 'current' for short). A current is a movement of electrons. Therefore, when current passes, (negative) charge is transferred (figure 2.8).

The amount of charge transferred is given by:
charge transferred = current × time

$$Q = It$$

where Q = charge transferred, I = current, t = time

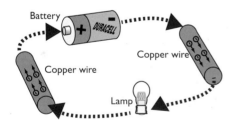

Figure 2.8 An electric current is the movement of electrons from the negative to the positive terminals of an energy source or battery.

Current is measured in amperes (A) and time in seconds (s) so 1 coulomb = 1 ampere second (1 C = 1 A s).

Example
There is a current of 4 A in a toaster when it is switched on. How much charge passes through the toaster when it is switched on for two minutes?

Solution

$Q = I\,t$
$Q = 4 \times (2 \times 60)$
$Q = 480\,C$

Example
A torch lamp passes 90 C of charge in 5 minutes. What is the current in the lamp?

Solution

$Q = It$
$90 = I \times (5 \times 60)$
$90 = 300\,I$
$I = \dfrac{90}{300} = 0.30\,A$

Conductors and insulators

Negative charges (i.e. electrons) can only move from the negative terminal to the positive terminal of a battery if there is an electrical path between them. Materials which allow negative charges to move through them easily, to form an electric current, are known as **conductors**. Conductors are mainly metals, such as copper, gold and silver. However, carbon is also a good conductor.

Materials which do not allow electrons to move through them easily are called **insulators**. Glass, plastic, wood and air are examples of insulators.

Voltage

Look again at figure 2.8. The battery changes chemical energy, from the substances inside it, into electrical energy. This electrical energy is carried by the charges (electrons) that move round the circuit and is given up as heat and light as they pass through the filament of the lamp.

The electrical energy given to the negative charges by the battery is a measure of the **voltage** of the battery. To be exact, the voltage of a battery is the electrical energy given to one coulomb of charge passing through the battery. For example, a 9 volt battery gives 9 joules of energy to each coulomb of charge passing through it. A 1.5 volt battery gives 1.5 joules of energy to each coulomb of charge passing through it.

Circuit symbols

Some commonly used circuit symbols are shown in figure 2.9.

(a)

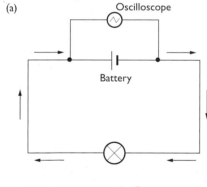

(b)

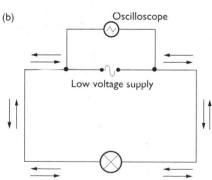

Figure 2.10 (a) Electrons move in only one direction. This is known as direct current (d.c.). (b) Electron movement is to and fro. This is known as alternating current (a.c.).

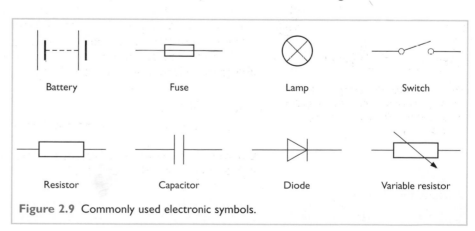

Figure 2.9 Commonly used electronic symbols.

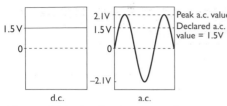

Figure 2.11 Oscilloscope traces from Figure 2.10(a) and (b).

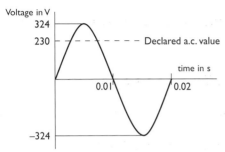

Figure 2.12 Oscilloscope traces for mains electricity.

Direct and alternating current

Figure 2.10(a) on page 33 shows a battery connected to a lamp and figure 2.10(b) shows a low voltage power supply connected to an identical lamp of equal brightness.

In figure 2.10(a) electrons (negative charges) move from the negative terminal through the lamp and wires to the positive terminal of the battery. This means that the electrons move in only one direction – this is known as **direct current** or. **d.c.** for short.

In figure 2.10(b) electrons move in one direction, then in the other direction and back again, that is the electrons move to and fro. This alternating movement of the electrons is known as **alternating current** or **a.c.** for short. The to and fro movement of the electrons is very frequent. It occurs 50 times every second and so the frequency of mains electricity is 50 hertz (50 Hz). The oscilloscope traces from figure 2.10 are shown in figure 2.11. The d.c. trace has a constant value of 1.5 V (volts) while the a.c. trace alternates from about +2.1 V to –2.1 V, a peak voltage of ±2.1 V. However, this a.c. voltage has a declared a.c. value of 1.5 V as it has the same effect on the lamp as the 1.5 V d.c. supply.

The declared a.c. voltage is always smaller than the peak a.c. voltage.

The alternating voltage of mains electricity in the UK is 230 volts – this is the declared a.c. value of the voltage (figure 2.12).

Section 2.2 Summary

◆ Mains supply is a.c. and a battery is d.c.
◆ Mains supply has a frequency of 50 Hz.
◆ The declared a.c. value of mains voltage is 230 V.
◆ The declared a.c. value of an alternating voltage is less than the peak value.
◆ Charge is measured in coulombs (C), current in amperes (A) and time in seconds (s).
◆ Charge = current × time, i.e. $Q = I\,t$.
◆ The voltage of a supply is a measure of the energy given to the charges in a circuit.

End of Section Questions

1 The current in a wire is 1.5 A. How much charge flows through the wire in 25 seconds?

2 A food mixer is switched on for 45 seconds. During this time 54 C of charge pass through it. Calculate the current in the food mixer.

3 A charge of 6500 C is transferred by a current of 2.5 A. How long did it take for the charge to be transferred?

4 Draw the circuit symbol for

(a) a battery; (b) a lamp:
(c) a switch; (d) a resistor;
(e) a variable resistor; (f) a fuse;
(g) a diode; (h) a capacitor.

5 A torch lamp is connected to a battery of voltage 1.5 V.

(a) Describe the movement of charge in the lit lamp.
(b) What does a voltage of 1.5 V mean?

6 A vacuum cleaner is connected to the mains a.c. supply.

(a) State the declared a.c. value, and the frequency of the mains supply.
(b) Describe the movement of charge in the vacuum cleaner when it is switched on.

7 An a.c. supply has a declared a.c. value of 10 V. Student A states that the peak value of this supply is 7 V, student B states that the peak value is 10 V and student C states that the peak value is 14 V. Which student is correct?

8 The diagram shows the rating plate of a vacuum cleaner.

1500 W
50 Hz ~

230 V
Serial No 130957

Figure 2.2Q8

(a) What value of voltage does it require to work properly?
(b) What is the frequency of the electrical supply to the vacuum cleaner?
(c) What value of fuse will be required to be fitted to the three-pin plug?
(d) How many cores will there be in the flex? Explain your answer.

At the end of this section you should be able to:

1 Draw and identify the circuit symbols for an ammeter and voltmeter.
2 Draw circuit diagrams to show the correct positions of ammeter and voltmeter in a circuit.
3 State that an increase in resistance of a circuit leads to a decrease in the current in that circuit.
4 Carry out calculations involving the relationship between resistance, current and voltage.
5 Use correctly the unit 'ohm'.
6 Give two practical uses of variable resistors.
7 State that when there is an electric current in a wire, there is an energy transformation.
8 Give three examples of resistive circuits in the home, in which electrical energy is transformed into heat.
9 State that the electrical energy transformed each second $= I\,V$.
10 State the relationship between energy and power.
11 Use correctly in context, the terms energy, power, joule and watt.
12 Carry out calculations involving the relationship between power, current and voltage.
13 State that in a lamp, electrical energy is transformed into heat and light.
14 State that the energy transformation in an electric lamp occurs in resistance wire (filament lamp) or gas (discharge tube).
15 State that a discharge tube lamp is more efficient than a filament lamp (i.e. more of the energy is transformed into light and less into heat).
16 State that the energy transformation in an electric heater occurs in resistance wire (element).
17 State that V/I for a resistor remains approximately constant for different currents.
18 Explain the equivalence of $I\,V$ and $I^2 R$.
19 Carry out calculations using the relationship between power, current and resistance.

Measuring current and voltage

Measuring current

Electric current is measured in amperes (A) and we use an **ammeter** to measure it. Figure 2.13 shows how an ammeter is connected in an electrical circuit.

An ammeter measures the current *through* a component.

Measuring voltage

Voltage is measured in volts (V) and we use a **voltmeter** to measure it. Figure 2.14 shows how a voltmeter is connected to an electrical circuit.

A voltmeter measures the voltage *across* a component.

Ammeters and voltmeters can be connected to the same circuit using the instructions given in figures 2.13 and 2.14.

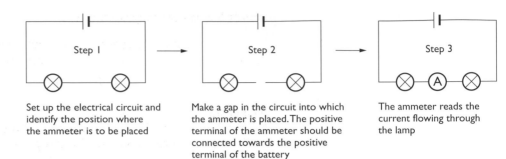

Step 1	Step 2	Step 3
Set up the electrical circuit and identify the position where the ammeter is to be placed	Make a gap in the circuit into which the ammeter is placed. The positive terminal of the ammeter should be connected towards the positive terminal of the battery	The ammeter reads the current flowing through the lamp

Figure 2.13 Connecting an ammeter into an electrical circuit.

(a)

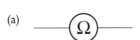

(b)

Figure 2.15 (a) The symbol for an ohmmeter, (b) measuring resistance using an ohmmeter.

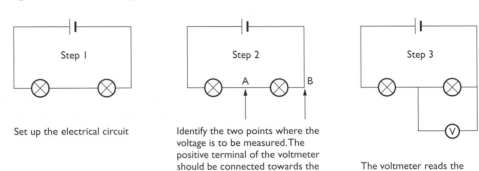

Step 1	Step 2	Step 3
Set up the electrical circuit	Identify the two points where the voltage is to be measured. The positive terminal of the voltmeter should be connected towards the positive terminal of the battery	The voltmeter reads the voltage across the lamp

Figure 2.14 Connecting a voltmeter into an electrical circuit.

Resistance

All materials oppose current passing through them. This opposition to the current is called **resistance**. Resistance is measured in **ohms** (Ω).

For most materials resistance depends on the:

◆ Type of material – the better the conductor the lower the resistance.
◆ Length of the material – the longer the material the higher the resistance.
◆ Thickness of the material – the thinner the material the higher the resistance.
◆ Temperature of the material – the higher the temperature the higher the resistance.

For a resistor, the resistance value remains constant for different currents, provided the temperature of the resistor does not change.

An **ohmmeter** can be used to measure resistance. The circuit symbol for an ohmmeter is shown in figure 2.15(a). Figure 2.15(b) shows an ohmmeter being used to measure the resistance of a resistor. The resistance of a resistor can also be measured using an ammeter and a voltmeter as shown in figure 2.16. In this method the voltage across, and the current through, the resistor have to be measured.

The resistance of the resistor can then be calculated using:

Figure 2.16 The resistance of the resistor (in this case a lamp) can be calculated using the voltmeter and ammeter readings.

$$\text{resistance of resistor} = \frac{\text{voltage across resistor}}{\text{current through resistor}}$$

That is
$$R = \frac{V}{I}$$

This is known as **Ohm's law**.

In units:

$$\text{ohms } (\Omega) = \frac{\text{volts (V)}}{\text{amperes (A)}}$$

Ohm's law is normally written as:

$$\begin{array}{c}\text{voltage across} \\ \text{resistor}\end{array} = \begin{array}{c}\text{current through} \\ \text{resistor}\end{array} \times \begin{array}{c}\text{resistance of} \\ \text{resistor}\end{array}$$

$$\text{or } V = I\,R$$

Example

A lamp has a voltage of 12 V across it and a current of 1.5 A passing through it. Calculate the resistance of the lamp.

Solution

$$V = I\,R$$
$$12 = 1.5\,R$$
$$R = \frac{12}{1.5} = 8.0\,\Omega$$

A resistor whose resistance can be changed is known as a **variable resistor**. The resistance is normally changed by altering the length of the wire in the resistor (the longer the wire, the higher the resistance). Variable resistors are often used as volume or brightness controls on televisions, and dimmers on lights.

Power

When an electric current passes through a wire, the electrons making up the current collide with the atoms of the wire. These collisions make the wire hotter and so some of electrical energy is changed into heat in the wire. The amount of heat produced depends on the current and the resistance of the wire. Heating elements for electric fires and kettles change electrical energy into heat energy in the resistive wire inside the element.

A lamp transfers electrical energy into heat energy and light energy in a resistance wire called the **filament**. How quickly it does this is known as the **power rating** of the lamp.

Power is the energy transferred in one second:

$$\text{Power} = \frac{\text{energy transferred}}{\text{time taken}}$$

$$P = \frac{E}{t}$$

In units:

$$\text{watts (W)} = \text{joule per second (J/s)}$$

$$1\,\text{W} = 1\,\text{J/s}$$

Example

A lamp uses 72 000 joules of energy in a time of 12 minutes. What is the power of the lamp?

Solution

$$P = \frac{E}{t} = \frac{72\,000}{12 \times 60} = 100\,W$$

Power, current and voltage

Four different lamps of known power ratings were connected to an electrical supply; the readings obtained for the voltage across and the current through the lamps are shown in table 2.3.

Power rating of lamp in W	Voltage in V	Current in A
6	12	0.5
24	12	2.0
36	12	3.0
48	12	4.0

Table 2.3

Calculate the value of current × voltage, for each lamp, and compare it with the power rating of the lamp. You should see that the power rating of the lamp is equal to the current multiplied by the voltage. From this we have:

$$\text{power} = \text{current} \times \text{voltage}$$

$$P = IV$$

But from Ohm's law $\quad V = IR$

$$P = I(I\,R)$$

$$P = I^2R$$

Alternatively:

$$P = IV$$

But from Ohm's law $\quad I = \dfrac{V}{R}$

$$P = \dfrac{(V)}{R}V$$

$$P = \dfrac{V^2}{R}$$

The equations $P = IV$, $P = I^2R$ and $P = \dfrac{V^2}{R}$ are used to find the power rating of appliances.

Use Ohm's Law to calculate the resistance of each of the lamps in the table and then check that the above equations can be used to calculate the power rating of the lamps.

Example
The interior light of a car is operated from a 12 V car battery. A current of 0.25 A passes through the lamp when it is switched on. Calculate the power rating of the lamp.
Solution

$$P = IV = 0.25 \times 12 = 3.0\,\text{W}$$

Example
A lamp has a resistance of 4 Ω and a current of 3 A passes through the lamp. Calculate the power rating of the lamp.
Solution

$$P = I^2\,R = 3^2 \times 4 = 36\,\text{W}$$

Filament lamps

Figure 2.17 A 20W compact fluorescent lamp (left), and some 100w filament lamps. Each fluorescent lamp gives out about the same amount of light as the filament lamp.

An electric lamp consists of a filament (usually of thin tungsten wire) housed in a glass container. As an electric current passes through the resistance wire (filament), so much electrical energy is changed into heat energy that the filament glows white hot. Filament lamps produce both heat and light. Electrical energy is changed into heat energy and light energy in the filament (resistance wire) of the lamp. Filament lamps are normally classified in terms of their power rating. The most common household lamps are 100 W (watt), 60 W and 40 W. They will work for approximately 1000 hours before the filament breaks (figure 2.17).

Gas discharge lamps

Gas discharge lamps depend on an electric current passing through a gas or vapour. They are cooler in operation as more of the electrical energy is converted into light energy. Discharge lamps are therefore much more efficient than filament lamps. There are a number of different types of discharge lamps such as neon, argon, mercury and sodium vapour lamps. Fluorescent lamps are filled with mercury vapour at low pressure. When an electric current is passed through the gas it produces invisible ultraviolet light. As the human eye would not see this, the lamps are coated on the inside with a special chemical which absorbs ultraviolet light and produces visible light. This process is called **fluorescence**.

Compact fluorescent lamps are available which provide the same amount of light as the filament lamps above, but which only use about one quarter of the electrical power. They also last about eight times as long as filament lamps but are more expensive to buy – although their whole-life cost is much lower than a filament lamp.

Section 2.3 Summary

- Voltage is measured in volts (V), current in amperes (A), resistance in ohms (Ω) and power in watts (W).
- Voltage across a resistor = current through resistor × resistance of resistor
  i.e. $V = IR$ – this is known as Ohm's law.
- The resistance of a resistor remains constant for different currents provided the temperature of the resistor does not change.
- An ammeter is connected in series and measures the current through a resistor.
- A voltmeter is connected in parallel and measures the voltage across a resistor.
- In a lamp, electrical energy is changed into heat and light in the resistance wire (filament).
- In a discharge lamp, more electrical energy is changed into light than in a filament lamp.
- Power = $\dfrac{\text{energy}}{\text{time}}$ i.e. $P = \dfrac{E}{t}$
- Power = current × voltage i.e. $P = IV$
- Power = current² × resistance i.e. $P = I^2 R$
- Power = $\dfrac{\text{voltage}^2}{\text{resistance}}$ i.e. $P = \dfrac{V^2}{R}$

1   Redraw the diagram shown below to show both a voltmeter connected to measure the voltage across component R and an ammeter connected to measure the current through component S.

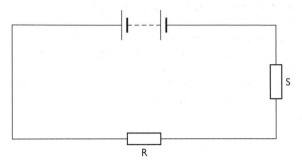

Figure 2.3Q1

2   A 150 Ω resistor has a current of 0.060 A passing through it. What is the voltage across the resistor?

3   A torch lamp is connected to a 6 V battery. The current in the lamp is 0.50 A. What is the resistance of the lamp?

4   A 47 Ω resistor is connected to a 12 V supply. What current passes through the resistor?

5   The element of an electric cooker operates from the 230 V mains supply. When it is switched on a current of 4.6 A passes through the element. Calculate the power rating of the element.

6   A television is rated at 276 W, 230 V. Calculate the current in the television when it is working at its correct rating.

7   A lamp is rated at 24 W, 2 A. Calculate the voltage across the lamp when it is working at its correct rating.

8   When operating, the heating element of a hairdryer has a resistance of 57.5 Ω. The current in the element is 4.0 A. Calculate the power rating of the element.

9   When switched on, the power rating of the heating element of a toaster is 1035 W. The current in the element is 4.5 A. Calculate the resistance of the element when it is operating.

10   An electric drill has a power rating of 460 W. When operating, the drill has a resistance of 115 Ω. Calculate the current passing through the drill.

11   A heating element is connected to a 12 V supply. A current of 1.5 A passes through the element. Find:
(a) the power rating of the element;
(b) the energy dissipated (used up) by the element in three minutes.

12   The heating element of a cooker has a resistance of 50 Ω. The element is connected to a 230 V supply. What is the power rating of the element?

13   A spotlight is rated at 50 W, 12 V. Calculate the resistance of the spotlight when operating at its correct rating.

At the end of this section you should be able to:

1 State a practical application in the home, which requires two or more switches used in series.
2 State that in a series circuit, the current is the same at all points.
3 State that the sum of currents in parallel branches is equal to the current drawn from the supply.
4 Explain that connecting too many appliances to one socket is dangerous because a large current could be drawn from the supply.
5 State that the voltage across components in parallel is the same for each component.
6 State that the sum of voltages across components in series is equal to the voltage of the supply.
7 Describe how to make a simple continuity tester.
8 Describe how a continuity tester may be used for fault finding.
9 Draw circuit diagrams to describe how the various car lighting requirements are achieved.
10 Carry out calculations involving the relationships $R_T = R_1 + R_2 + \ldots$ and $\dfrac{1}{R_T} = \dfrac{1}{R_1} + \dfrac{1}{R_2} + \ldots$

Types of circuit

Electrical components, such as lamps and resistors, can be connected in series, in parallel or a mixture of series and parallel. A **series circuit** has only one electrical path from the negative terminal of the battery to the positive terminal. A **parallel circuit** has more than one electrical path, known as branches, from the negative terminal of the battery to the positive terminal. Figure 2.18(a) shows three lamps connected in series while figure 2.18(b) shows three lamps connected in parallel. Figure 2.18(c) shows a mixed series and parallel circuit in which two resistors are connected in parallel and a lamp is connected in series.

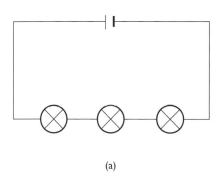
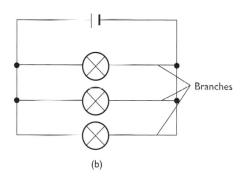
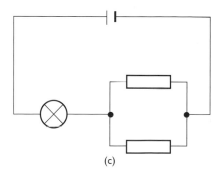

Figure 2.18 (a) A series circuit, (b) a parallel circuit, (c) a mixed series and parallel circuit.

Figure 2.19 An ohmmeter measures the combined resistance of three resistors connected in series.

A series circuit

Three resistors of value $1\,\Omega$, $4\,\Omega$ and $3\,\Omega$ are connected in series. Their combined (or total) resistance is measured by an ohmmeter as $8\,\Omega$, as shown in figure 2.19.

Figure 2.20(a) shows these same three resistors connected in series but with ammeters connected to measure the current at various positions. Voltmeters have also been connected to measure the voltage across each resistor. Table 2.4 shows the readings on the meters. Use Ohm's law, to confirm the value of each resistor.

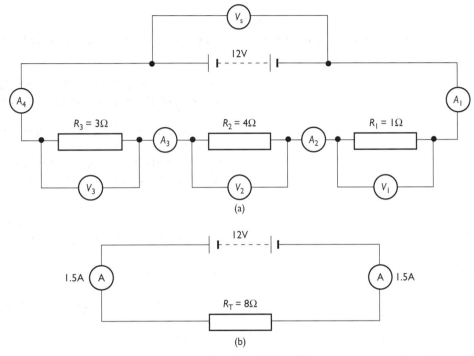

Figure 2.20 (a) Measuring the current through and the voltage across resistors in series, (b) the equivalent circuit to that shown in figure (a).

Voltage in V	Current in A	Resistance in Ω
$V_1 = 1.5$	$A_1 = 1.5$	$R_1 = 1$
$V_2 = 6.0$	$A_2 = 1.5$	$R_2 = 4$
$V_3 = 4.5$	$A_3 = 1.5$	$R_3 = 3$
$V_S = 12$	$A_4 = 1.5$	$R_T = 8$

Table 2.4

◆ What do you notice about the current at different positions in a series circuit?
◆ What do you notice about the voltages V_1, V_2, V_3 and V_S for a series circuit?
◆ What do you notice about the resistances R_1, R_2, R_3 and R_T for a series circuit?

The circuit in figure 2.20(b) is equivalent to that in (a), that is the two circuits are the same, as both have the same supply voltage and the same current drawn from the supply.

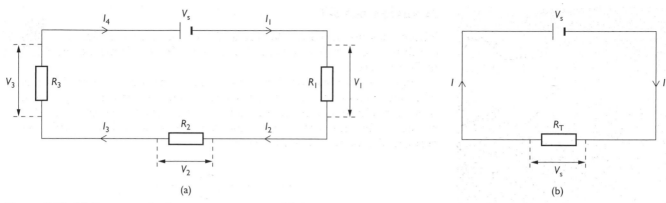

Figure 2.21 (a) Current and voltage in a series circuit, (b) the equivalent circuit to (a) provided $R_T = R_1 + R_2 + R_3$.

In the series circuit shown in figure 2.21(a):

♦ The current is the same at all positions – the current does not split up, i.e. $I_1 = I_2 = I_3 = I_4$.
♦ The supply voltage is equal to the sum of the voltages round the circuit, i.e. $V_S = V_1 + V_2 + V_3$.
♦ The total resistance (R_T) of the circuit is equal to the sum of the individual resistances, i.e. $R_T = R_1 + R_2 + R_3$.

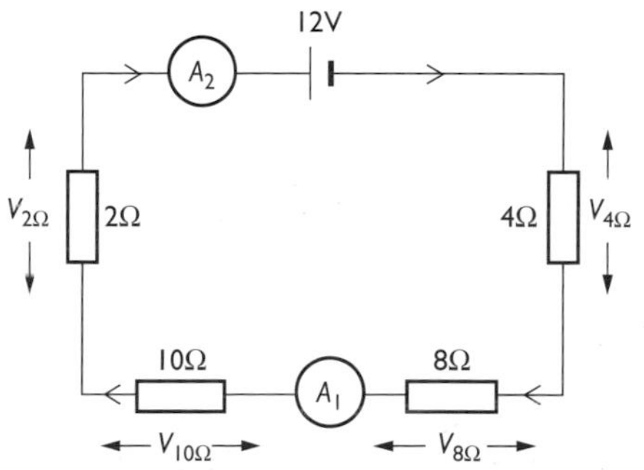

Example
For the circuit shown (figure 2.22), calculate the readings on the ammeters A_1 and A_2, and the voltage across each of the resistors.

Solution
Total resistance of circuit:

$$R_T = R_1 + R_2 + R_3 + R_4 = 4 + 8 + 10 + 2 = 24\,\Omega$$

From Ohm's law:

$$\text{circuit current} = I = \frac{V_S}{R_T} = \frac{12}{24} = 0.5\,\text{A}$$

So ammeter $A_1 = A_2 = 0.5\,\text{A}$ (since current in a series circuit is the same at all points). The voltages across each resistor are calculated using Ohm's law:

$$\text{voltage across resistor} = \text{current through resistor} \times \text{resistance of resistor}$$

$$
\begin{aligned}
V_{4\Omega} &= I \times R_{4\Omega} = 0.5 \times 4 = 2.0\,\text{V} \\
V_{8\Omega} &= I \times R_{8\Omega} = 0.5 \times 8 = 4.0\,\text{V} \\
V_{10\Omega} &= I \times R_{10\Omega} = 0.5 \times 10 = 5.0\,\text{V} \\
V_{2\Omega} &= I \times R_{2\Omega} = 0.5 \times 2 = 1.0\,\text{V}
\end{aligned}
$$

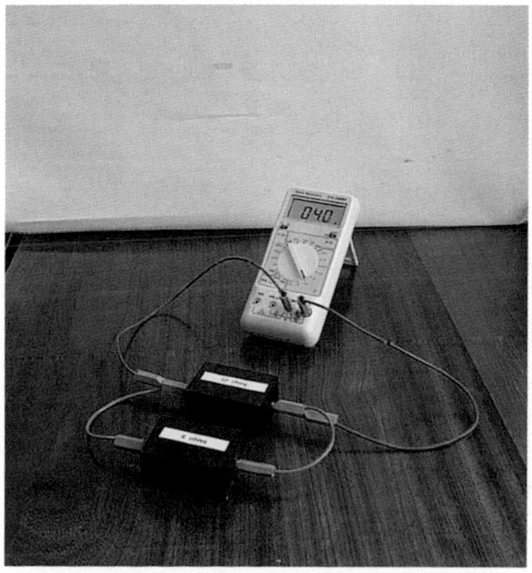

Figure 2.22 Example for a series circuit.

A parallel circuit

Two resistors of value $6\,\Omega$ and $12\,\Omega$ are connected in parallel. Their combined (or total) resistance is measured by an ohmmeter as $4\,\Omega$, as shown in figure 2.23.

Figure 2.24(a) shows these resistors connected in parallel with ammeters connected to measure the current at various positions. Voltmeters have also been connected to measure the voltage across each resistor. Table 2.5 shows the readings on the meters. Use Ohm's law to confirm the value of each resistor.

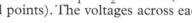

Figure 2.23 An ohmmeter measures the combined resistance of two resistors connected in parallel.

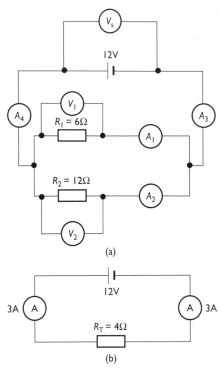

Voltage in V	Current in A	Resistance in Ω
$V_1 = 12$	$A_1 = 2$	$R_1 = 6$
$V_2 = 12$	$A_2 = 1$	$R_2 = 12$
–	$A_3 = 3$	–
–	$A_4 = 3$	–
$V_S = 12$	–	$R_T = 4$

Table 2.5

♦ What do you notice about the voltages across resistors connected in parallel?
♦ What do you notice about the ammeter readings A_1, A_2, A_3 and A_4 for a parallel circuit?
♦ What do you notice about the sizes of the resistances R_1 and R_2 compared with R_T?

The circuit in figure 2.24(b) is equivalent to that in (a), that is the two circuits are the same, as both have the same supply voltage and the same current drawn from the supply.

The combined resistance of $4\,\Omega$ is obtained from the two resistors as follows:

$$\frac{1}{R_T} = \frac{1}{R_1} + \frac{1}{R_2} = \frac{1}{6} + \frac{1}{12} = 0.167 + 0.083 = 0.25$$

$$\frac{1}{R_T} = 0.25$$

$$R_T = \frac{1}{0.25} = 4\,\Omega$$

Notice that this can also be obtained using Ohm's law in the equivalent circuit shown in figure 2.23(b).

$$V_S = I\,R$$
$$12 = 3\,R$$
$$R = \frac{12}{3} = 4\,\Omega$$

For the parallel circuit shown in figure 2.25:
♦ The current splits up so the circuit current equals the sum of the currents in the branches, i.e. $I = I_1 + I_2$.
♦ The voltage across resistors connected in parallel is the same, i.e. $V_1 = V_2$ and in this case $V_S = V_1 = V_2$.
♦ The total resistance of the circuit is found using: $\frac{1}{R_T} = \frac{1}{R_1} + \frac{1}{R_2}$.

Note: For a parallel circuit, the total resistance is less than the value of the smallest resistance.

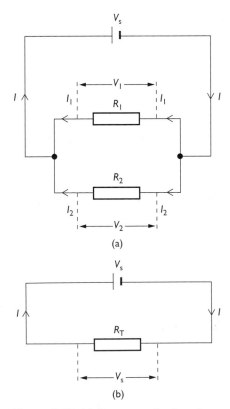

Figure 2.24 (a) Measuring the current through and the voltage across resistors in parallel, (b) the equivalent circuit to that shown in figure (a).

Figure 2.25 (a) Current and voltage in a parallel circuit, (b) the equivalent circuit to (a) provided $1/R_T = 1/R_1 + 1/R_2$.

Example

Three resistors of resistance $100\,\Omega$, $200\,\Omega$ and $100\,\Omega$ are connected in parallel. Calculate their total resistance.

Solution

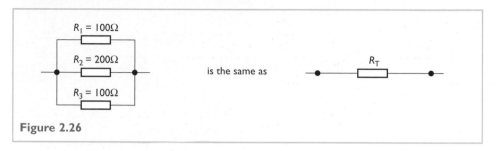

Figure 2.26

$$\frac{1}{R_T} = \frac{1}{R_1} + \frac{1}{R_2} + \frac{1}{R_3} = \frac{1}{100} + \frac{1}{200} + \frac{1}{100} = 0.01 + 0.005 + 0.01 = 0.025$$

$$\frac{1}{R_T} = 0.025$$

$$\frac{1}{R_T} = \frac{1}{0.025} = 40\,\Omega$$

As a check on our answer we would expect it to be smaller than $100\,\Omega$.

Example

Two resistors each of resistance $20\,\Omega$ are connected in parallel. Calculate their total resistance.

Solution

$$\frac{1}{R_T} = \frac{1}{R_1} + \frac{1}{R_2} = \frac{1}{20} + \frac{1}{20} = 0.05 + 0.05$$

$$\frac{1}{R_T} = 0.10$$

$$R_T = \frac{1}{0.10} = 10\,\Omega$$

Note: The total resistance of two identical resistors connected in parallel is half that of one of the resistors.

Example

Three resistors are connected to a 12 V supply as shown in the circuit in figure 2.27. Calculate:
(a) the total resistance of the circuit;
(b) the current in the 10 Ω resistor;
(c) the voltage across the 25 Ω resistor.

Solution

(a)

$$\frac{1}{R_{AB}} = \frac{1}{R_1} + \frac{1}{R_2} = \frac{1}{100} + \frac{1}{25} = 0.01 + 0.04 = 0.05$$

$$R_{AB} = \frac{1}{0.05} = 20\,\Omega$$

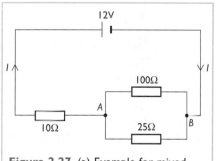

Figure 2.27 (a) Example for mixed series and parallel circuit.

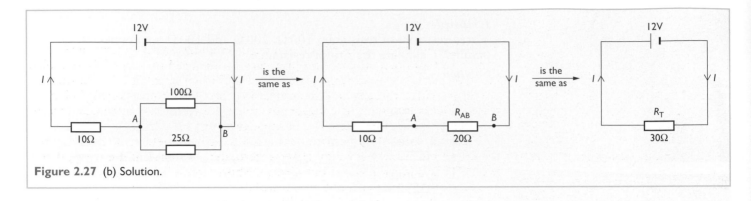

Figure 2.27 (b) Solution.

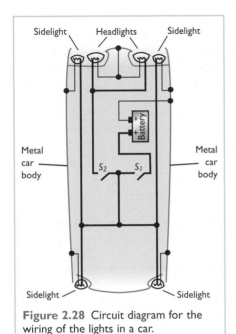

Figure 2.28 Circuit diagram for the wiring of the lights in a car.

Fascinating Physics

See figure 2.27(b).

$$\text{Total resistance } R_T = 10 + R_{AB} = 10 + 20 = 30\,\Omega$$

(b)

$$V_S = I\,R_T \text{ gives } 12 = I \times 30, \text{ hence } I = \frac{12}{30} = 0.4\,\text{A}$$

(c)

Voltage across $10\,\Omega$ resistor = current through resistor × resistance of resistor

$$V = I\,R = 0.4 \times 10 = 4\,V$$

$$\text{But supply voltage} = V_{10\,\Omega} + V_{AB}$$

$$12 = 4 + V_{AB}$$

$$V_{AB} = 8\,V$$

i.e. Voltage across $25\,\Omega$ resistor = V_{AB} since voltage across resistors connected in parallel is the same

$$= 8\,V$$

Superconductors

At room temperature, all materials have resistance. As most materials are cooled to lower and lower temperatures, their resistance decreases. However, the electrical resistance of some metals and alloys when at very low temperatures of about –270°C becomes zero. These materials are known as superconductors. Superconductors are ideal in applications where you don't want electrical energy lost as heat. This is put to use in very strong electromagnets where an intense magnetic field is required.

A practical parallel circuit – car wiring

Figure 2.28 shows how the sidelights and headlights in a car are wired. When S_1 is closed the sidelights come on. When S_1 and S_2 are both closed, the headlights and the sidelights are both on. The lights are wired in parallel so that (a) all lamps have the same voltage and (b) if one lamp breaks then the others continue to light.

Figure 2.29 (a) A short circuit, (b) an open circuit.

Fault finding

There are two types of fault that can occur in an electrical circuit – a short circuit and an open circuit. An ohmmeter can be used to test for these faults. However, an ohmmeter can only be used to find an open circuit or a short circuit provided there is no current passing through the component being tested. The component or components must therefore be disconnected from a power supply before using the ohmmeter to test for a fault.

Figure 2.29(a) shows a wire connected to an ohmmeter. The reading on the ohmmeter is zero (or very close to zero, the resistance of the wire), that is $0\,\Omega$. This is known as a **short circuit**. Short circuits can only occur with an electrical component, not with a wire.

Figure 2.29(b) shows a wire with an electrical break connected to the ohmmeter. The reading on the ohmmeter is the largest value the meter can indicate (on most models of ohmmeter this is shown by the number '1' on the left hand side of the display), that is infinity. This is known as an **open circuit**. Open circuits can occur with both electrical components and wires. Consider a circuit in which a lamp does not light. This could be due to:

◆ the filament of the lamp being broken, i.e. an open circuit lamp (a likely possibility);
◆ a piece of wire being connected in parallel with the lamp, i.e. a short-circuited lamp;
◆ a broken wire between the battery and the lamp;
◆ the battery being 'flat'.

Using the ohmmeter, the fault in this circuit may be found by:
(a) Testing the lamp. When the ohmmeter is connected across the lamp, it will show:
◆ a reading which is the same as the resistance of a working lamp;
◆ a very, very high reading, i.e. open circuit, indicating that the lamp filament is broken (fault found);
◆ zero resistance, i.e. short circuit (in this case a wire is connected in parallel with the lamp) (fault found).
(b) Testing the wires in the circuit. Each wire is connected to the ohmmeter as shown in figure 2.29(a). It will show:
◆ zero resistance, i.e. short circuit, the wire provides an electrical path;
◆ a very, very high reading, i.e. an open circuit, indicating that the wire is broken (fault found).

If none of the above has detected a fault then the battery should be replaced.

A simple circuit tester

An ohmmeter can be an expensive piece of equipment. A cheaper and simpler circuit tester is shown in figure 2.30. The resistor R is present so that when wires X and Y are joined, the lamp is as bright as it safely can be – it is called a **protective resistance** and prevents too much current passing through the lamp and 'blowing' it. Any other piece of electrical equipment, for example a resistor or a lamp joined between X and Y will increase the resistance of the circuit and so there will be less current and the lamp will be less bright.

Before using the tester to detect for a fault, X and Y are joined together to check that the lamp lights. If it does not light, the lamp or battery should be replaced – after having checked for a loose connection.

Figure 2.30 A simple circuit tester.

◆ In a series circuit:
  - the current is the same at all points, i.e. $I_1 = I_2 = I_3$;
  - the supply voltage is equal to the sum of the voltages across components, i.e. $V_S = V_1 + V_2 + V_3$;
  - the total (or combined) resistance is equal to the sum of the individual resistors i.e. $R_T = R_1 + R_2 + R_3$.

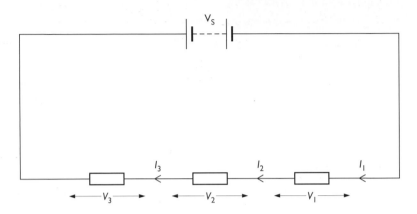

◆ In a parallel circuit:
  - the circuit current is equal to the sum of the currents in the branches, i.e. $I_C = I_1 + I_2 + I_3$;
  - the voltage across components is the same, i.e. $V_S = V_1 = V_2 = V_3$;
  - the total (or combined) resistance is given by $\dfrac{1}{R_T} = \dfrac{1}{R_1} + \dfrac{1}{R_2} + \dfrac{1}{R_3}$

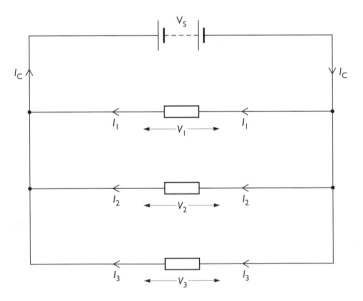

◆ In a parallel circuit, the total resistance is less than the resistance of the smallest resistor.
◆ An open circuit has a resistance which is so large that it cannot be measured, i.e. infinite resistance (no current can pass).
◆ A short circuit has a very, very small resistance.

End of Section Questions

1. In the circuits shown in the figure, state the missing readings on ammeters A_1, A_2 and A_3, and the readings on voltmeters V_1, V_2 and V_3.

(a)

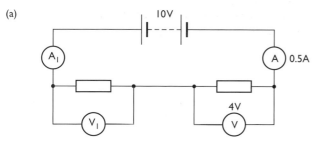

(b)

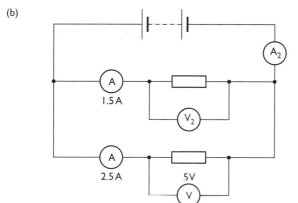

(c)

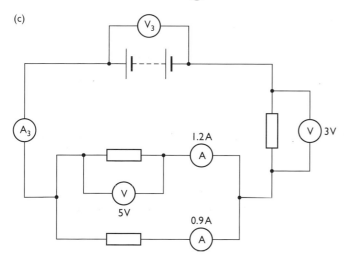

Figure 2.4Q1

2. Find the total resistance between A and B in the circuits shown in the figure.

(a)

(b)

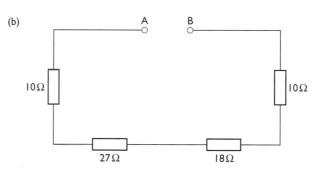

Figure 2.4Q2

3. Find the total resistance between C and D in the circuits shown in the figure.

(a)

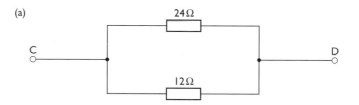

(b)

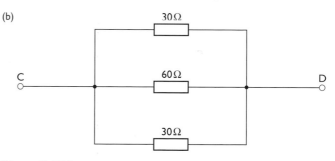

Figure 2.4Q3

4 Find the total resistance between E and F in the circuits shown in the figure.

(a)

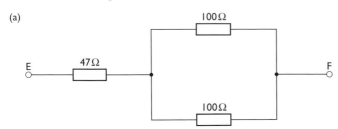

(b)

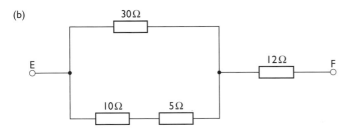

Figure 2.4Q4

5 Three resistors, of value 6 Ω, 10 Ω and 8 Ω, are connected in series to a 6 V supply. Calculate the:

(a) total resistance of the circuit;
(b) current passing through each resistor.

6 Two resistors, of value 30 Ω and 60 Ω, are connected in parallel to a 10 V supply. Calculate the:

(a) total resistance of the circuit;
(b) current drawn from the supply.

7 A 16 V supply, a lamp and a resistor are connected in series. The lamp is operating at its correct rating of 24 W, 12 V.

(a) What is the voltage across the lamp?
(b) Find the current drawn from the supply.
(c) Calculate the resistance of the resistor.

8 One of the lamps in the circuit shown in the figure below is not working.

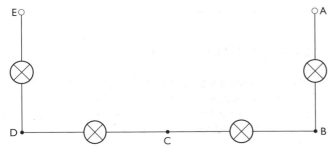

Figure 2.4Q8

(a) Describe how would you use an ohmmeter to discover which of the lamps was not working.
(b) Would the ohmmeter indicate a short circuit or an open circuit when connected across the broken lamp?

9 Three resistors of value 15 Ω, 30 Ω and 15 Ω are connected in series to the 230 V mains supply. Calculate the:

(a) total resistance of the circuit;
(b) current drawn from the supply;
(c) voltage across the 30 Ω resistor.

10 For the circuit shown in the figure calculate the:

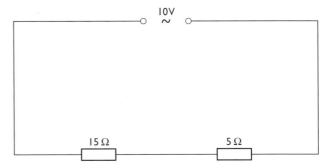

Figure 2.4Q10

(a) total resistance of the circuit;
(b) current through each resistor.

11 For the circuit shown in the figure calculate the:

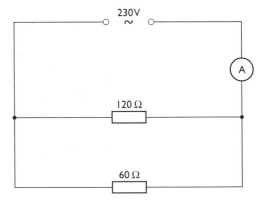

Figure 2.4Q11

(a) total resistance of the circuit;
(b) ammeter reading.

12 For the circuit shown in the figure calculate the:

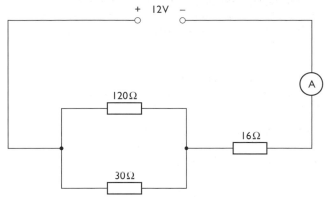

Figure 2.4Q12

(a) total resistance of the circuit;
(b) ammeter reading;
(c) voltage across the 30 Ω resistor.

13 What is the total resistance of three identical resistors of resistance 30 Ω connected in parallel?

SECTION 2.5   Behind the wall

At the end of this section you should be able to:

1 State that household wiring connects appliances in parallel.
2 State that mains fuses protect the mains wiring.
3 State that a circuit breaker is an automatic switch which can be used instead of a fuse.
4 State that kWh is a unit of energy.
5 Describe, using a circuit diagram, a ring circuit.
6 State advantages of using the ring circuit as a preferred method of wiring in parallel.
7 Give two differences between the lighting circuit and the power ring circuit.
8 State one reason why a circuit breaker may be used in preference to a fuse.
9 Explain the relationship between kilowatt-hours and joules.

Electricity and house wiring

Electricity is conveyed from power stations to our homes by the National Grid System (see Chapter 6). It arrives in our homes by a cable called the **service cable**. When the service cable enters your home various connections are made to it. Look at the lighting and ring main circuits shown in figure 2.31.

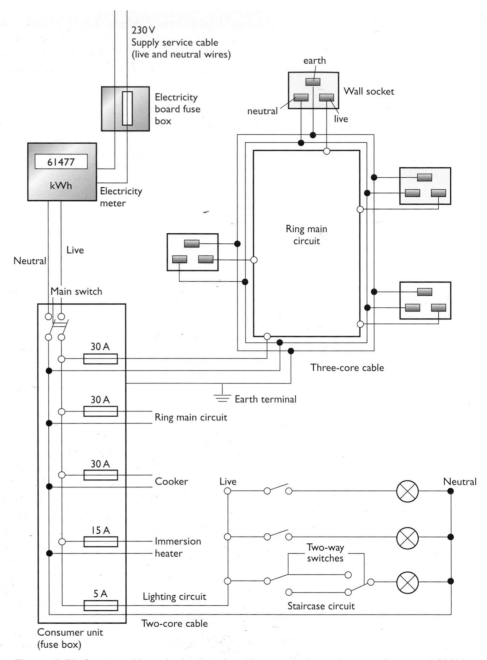

Figure 2.31 Service cable – electricity is brought into the house by an underground 230 V service cable (live and neutral wires).

◆ The **lighting circuit** – the house lights are connected in parallel with each other. Normally there are two lighting circuits (one for downstairs, the other for upstairs), each having a 5 A fuse in the **consumer unit** (or fuse box).

◆ The **ring main circuit** – the electrical sockets are connected in parallel to the live, neutral and earth wires which loop round the house.

There are normally two ring circuits (again one for downstairs, the other for upstairs), each having a 30 A fuse in the consumer unit. Notice that the ring main circuit is different from the lighting circuit in that the live, neutral and earth wires form a 'ring'.

The difference in these two circuits can be demonstrated in the 'conventional' parallel and ring circuits shown in figure 2.32. Both circuits have identical lamps of equal brightness and ammeters have been positioned in the circuits. The ammeter readings are shown in table 2.6.

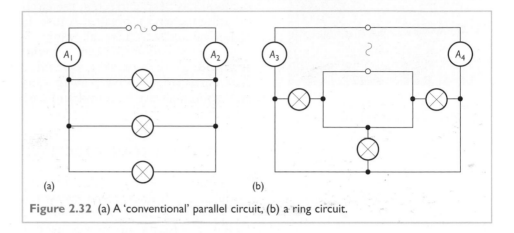

Figure 2.32 (a) A 'conventional' parallel circuit, (b) a ring circuit.

Circuit	Ammeter	Readings
Parallel	$A_1 = 1.8\,A$	$A_2 = 1.8\,A$
Ring	$A_3 = 0.9\,A$	$A_4 = 0.9\,A$

Table 2.6

Notice that the current in the ring circuit cables is half that in the parallel circuit cables. This is due to the current in a ring having two possible routes round the circuit. This means that although the power delivered to the lamps is the same for both circuits, the cables in a ring circuit carry a smaller current than the parallel circuit. Since the ring main cable carries less current, the cables will produce less heat. This is a safer arrangement and the cable can be made thinner. Thinner cable is less costly since it contains less conductor (usually copper).

A ring main is used where the power requirement can be high. The much lower power requirements of a lighting circuit already allow thin cables to be used and so a ring circuit in this case is not justified.

Mains fuses and circuit breakers

Domestic wiring circuits are protected by fuses in the consumer unit. These are in addition to the fuse in each three-pin plug. They are used to protect the hidden cables (behind the walls) from overheating. A fuse will melt and break the electrical circuit if the current becomes too high.

Mains fuses will 'blow' if the circuit is overloaded or if the fuse wire is of too low a rating. Typical values for consumer unit fuses are shown in table 2.7.

Circuit/Appliance	Fuse
lighting circuit	5 A
ring circuit	30 A
immersion heater	15 A
cooker	30 A

Table 2.7

Figure 2.33 Most modern house consumer units are fitted with circuit breakers.

Circuit breakers

Figure 2.33 shows a consumer unit fitted with circuit breakers. **Circuit breakers** are used in some consumer units instead of fuses. These automatically switch themselves off (trip) if the circuit is overloaded. When the fault has been corrected, the circuit can be reconnected simply by resetting the circuit breaker. A fuse, however, could be replaced with a higher value of fuse wire, allowing a faulty circuit to work. This very dangerous situation (the cables could overheat and catch fire!) cannot arise with a circuit breaker, as it will continue to trip until the fault has been corrected. Before replacing a fuse or resetting a circuit breaker, the fault should always be found and corrected.

Calculating your electricity bill

The electricity meter measures the electrical energy used by the appliances in your house. It does this in **kilowatt-hours (kWh)** or 'units' of electricity.

1 kWh represents the energy used by a 1 kW heater for one hour.

$$
\begin{aligned}
\text{Energy used} &= \text{power} \times \text{time} \\
&= 1\,\text{kW} \times 1\,\text{hour} \\
&= 1000\,\text{W} \times (60 \times 60)\,\text{seconds} \\
&= 3\,600\,000\,\text{Joules} \\
&= 3.6\,\text{MJ} \\
\text{i.e. } 1\,\text{kWh} &= 3.6\,\text{MJ}
\end{aligned}
$$

Number of kWh = number of kilowatts × number of hours
The more kWh or 'units' of electricity that are used in your house, the higher the cost of your electricity bill i.e.
Cost = number of kWh used × price of 1 kWh

Example
A 100 W lamp in a room is left on for six hours. Calculate the energy used by the lamp in kWh.
Solution

$$
\begin{aligned}
\text{Number of kWh} &= \text{number of kW} \times \text{number of h} \\
&= 0.1 \times 6 \\
&= 0.6\,\text{kWh}
\end{aligned}
$$

Example
A cooker element has a power rating of 1.5 kW and is switched on for 20 minutes. The householder pays 8 p for each unit of electricity used. Calculate the cost of the energy used by the cooker.
Solution

$$
\begin{aligned}
\text{Number of kWh} &= 1.5 \times \frac{20}{60} \\
&= 0.5\,\text{kWh} \\
\text{Cost} &= \text{number of kWh} \times \text{price of 1 kWh} \\
\text{Cost} &= 0.5 \times 8 \\
\text{Cost} &= 4\,\text{p}
\end{aligned}
$$

End of Section Questions

1   Give two differences between the lighting circuit and ring main circuit in a house.

2   The maximum current that a ring circuit can carry is greater than the maximum current a lighting circuit can carry. Which circuit, if any, uses thicker cable?

3   A 1 kW electric fire is switched on for 3 hours. How many kWh of electricity are used?

4   A kettle of power rating 2 kW is switched on 10 times in a day. It is on for three minutes on each occasion. How many kWh of electricity are used in a day?

5   Diagrams A and B in the figure opposite show identical appliances (represented by resistors) connected to identical alternating power supplies. The current through each appliance (resistor) is 4.0 A.

(a) State the readings on ammeters A_1 and A_2.
(b) State the readings on ammeters A_3 and A_4.
(c) Which diagram, A or B, represents a ring circuit?

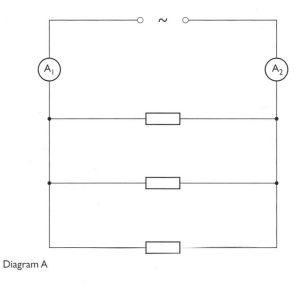

Diagram A

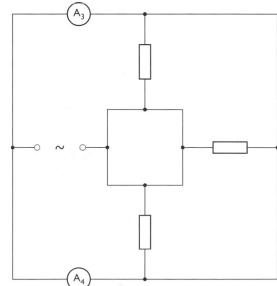

Diagram B

Figure 2.5Q5

(a)

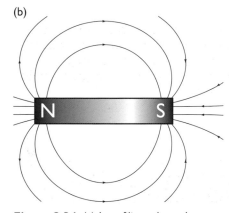

(b)

Figure 2.34 (a) Iron filings show the magnetic field lines, (b) direction of magnetic field lines for a permanent magnet.

Permanent magnets

A **permanent magnet**, as its name implies, has a magnetic field surrounding it which cannot be switched off. The opposite ends, or **poles**, of a magnet are called **north** and **south** (a north pole means a north-seeking pole, that is it always wants to point north). The shape of the magnetic field surrounding a magnet can be shown by scattering iron filings on a piece of paper placed on top of it. The direction of the magnetic field can be found using a compass (figure 2.34).

When two permanent magnets are placed close together, their magnetic fields produce forces in such a way that:
◆ a north pole repels a north pole;
◆ a south pole repels a south pole;
◆ and a north pole attracts a south pole.

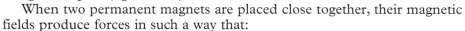

Like poles repel and unlike poles attract.

Some metals such as iron and steel are attracted to magnets.

Electromagnetism

Figure 2.35(a) shows a long straight wire passing vertically through a piece of card. A magnetic field surrounds the wire when it carries an electrical current. Increasing the current through the wire increases the strength of magnetic field surrounding the wire. Reversing the direction of the current through the wire reverses the direction of magnetic field around the wire. Figure 2.35(b) shows the pattern of magnetic field lines surrounding the wire when looking from above. Figure 2.36(a) shows the symbol which is used to indicate current passing out of the paper (towards you) and figure 2.36(b) shows the symbol for current passing into the paper (away from you).

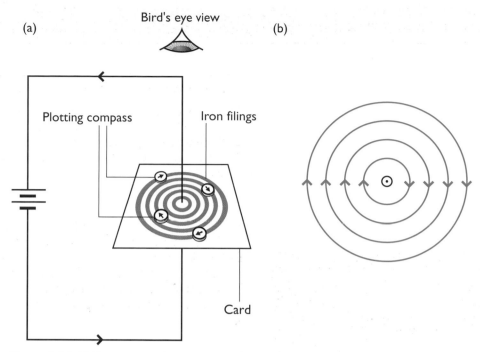

Figure 2.35 (a) The magnetic field surrounding a current-carrying wire, (b) the magnetic field lines as seen from above.

Figure 2.36 (a) Current passing out of paper, (b) current passing into paper.

Electromagnets

When an electric current passes through a wire which is coiled around an iron core, the core becomes magnetised and an **electromagnet** is produced, as shown in figure 2.37. However, the electromagnet has little strength without the iron core. The iron core is able to concentrate the magnetic field within itself, so giving a stronger magnetic effect.

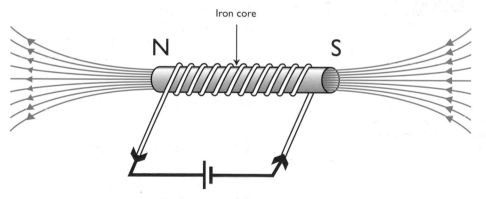

Figure 2.37 An electromagnet.

The magnetic field of an electromagnet can be made stronger by:
(a) increasing the current through the coils of wire;
(b) increasing the number of turns of wire on the core.
When an a.c. supply is used the electromagnet still produces a magnetic field – but one which alternates each time the current changes direction. There is no magnetic field when the electric current is switched off. This on-off nature of the magnetic field can be used in various ways, such as to lift (magnetic field) and release (no magnetic field) scrap iron as shown in figure 2.38.
Electromagnets are essential parts of many electrical devices.

Figure 2.38 Electromagnets can lift heavy objects.

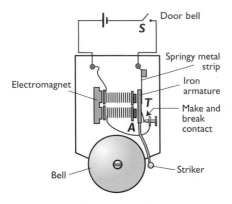
Figure 2.39 An electric bell.

The electric bell

Figure 2.39 shows an electric bell. When the door bell is pressed, switch S closes and completes the electrical circuit. A current passes through the electromagnet which becomes magnetised and attracts the iron bar or armature. However, when the striker moves to hit the bell, the electrical circuit is broken at T. There is no longer a current in the electromagnet and the electromagnet loses its magnetism. The springy metal strip is now able to pull the armature back, and in so doing the electrical circuit is re-made at T. This completes the circuit again and continuous ringing occurs as long as the door bell is pressed.

The magnetic relay

Figure 2.40 shows an electrically operated switch called a magnetic **relay**. When switch S is closed, there is a small current in the electromagnet. The magnetic field produced pulls the pivoted armature towards the iron core of the electromagnet, pushing the contacts closed. This completes the electrical circuit for the motor so turning it on. When switch S is opened, the electromagnet loses its magnetism, releasing the armature. The contacts open and the motor stops.

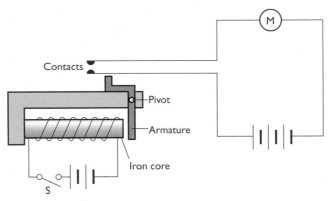
Figure 2.40 The motor is switched via the relay when switch S is closed.

Relays are very useful devices. They are used when it is desirable for a small current in a control circuit to open or close a switch in a main circuit which will carry a large current.

A current-carrying conductor in a magnetic field

Figure 2.41 shows a strip of aluminium foil which has been placed between the poles of a strong permanent magnet. When a current passes through the foil the magnetic field produced combines with the magnetic field of the permanent magnet to produce a force on the aluminium foil. The foil is pushed out of the field of the magnet. Figure 2.42 shows the same effect on current-carrying wires placed in the magnetic field of a permanent magnet. Notice that the force acting on the current-carrying wire is at right angles to the direction of the current and at right angles to the direction of the magnetic field. The direction of the force (or movement) on the wires can be reversed by either reversing the direction of the current, (i.e. reversing the terminals on the battery), or reversing the direction of the magnetic field (i.e. reversing the poles of the permanent magnet).

Figure 2.41 The current-carrying foil is forced out of the magnetic field of the permanent magnet.

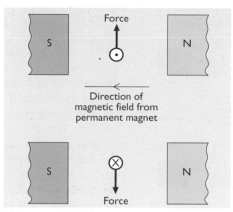

Figure 2.42 The force acting on a current-carrying conductor, when placed in an external magnetic field.

The force between a permanent magnet and a current-carrying conductor is put to use in an electric motor.

Trains, planes and magnets?

Magnetically levitating (maglev) trains 'float' above the track and can travel much faster than conventional trains. The duration of long-distance journeys, as a result, could be reduced by a third. High performance permanent magnets fixed on the underside of the carriage interact with a magnetic field created by an electromagnet positioned in the track. This produces levitation.

The electric motor

Figure 2.43 shows a simple model electric motor. The main parts of a motor are:

◆ a magnet;
◆ a few coils of wire (the rotating coil);
◆ split-ring commutator;
◆ carbon brushes;
◆ power supply.

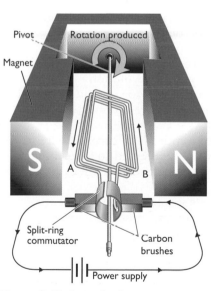

Figure 2.43 A simple electric motor.

A strong magnetic field exists between the poles of the magnet. The coils of wire are connected to the **split-ring commutator**. This assembly is free to rotate about a central axis within the magnetic field of the magnet. **Carbon brushes** are pushed lightly against the split-ring commutator by springs. *The carbon brushes allow current to pass from the power supply to the coils.* The current path is from the negative terminal of the supply through one brush, one half of the commutator, round the coils of wire, through the other half of the commutator, and on through the second brush to the positive terminal.

Due to the combined effect of the magnetic fields from the magnet and the current-carrying coils of wire, each side experiences a force, but in opposite directions as shown in figures 2.44(a) and (b). These make the coils turn clockwise, until it reaches a point where there is no current in the coils of wire and so there are no forces on the coils, figure 2.44(c). However, in practice the coils are carried slightly past this position. The current through the coils is now reversed by the commutator. The forces again have a clockwise turning effect since the current has been reversed, figures 2.44(d) and (e). The coils have now completed half a revolution. This is exactly the same situation as shown in figure 2.44(a) and so the process is repeated and the coils rotate. *The commutator automatically changes the direction of the current through the coils of wire every half revolution to give a continuous clockwise turning force.*

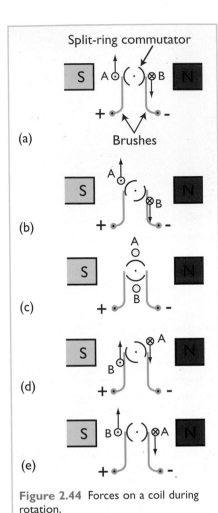

Split-ring commutator

Brushes

(a)

(b)

(c)

(d)

(e)

Figure 2.44 Forces on a coil during rotation.

This simple motor produces a low turning effect and a jerky action. Practical motors give a much improved performance because:

◆ Each coil consists of hundreds of turns of wire creating a greater turning force.
◆ Several coils are used, each set at a different angle and each connected to its own pair of commutator segments. This gives a greater turning force and smoother running.
◆ An electromagnet, called the **field coil**, replaces the permanent magnet. An electromagnet is more compact and produces a stronger magnetic field than a permanent magnet of the same size.

In practical motors the brushes are made of carbon because it:

◆ gives good electrical contact;
◆ moulds itself to the shape of the commutator;
◆ withstands high temperature;
◆ reduces wear on the commutator.

Section 2.6 Summary

◆ An electric motor consists of a rotating coil, field coil (magnet), brushes and a commutator.
◆ A magnetic field surrounds a current-carrying wire.
◆ The magnetic field surrounding an electromagnet allows many electrical devices to operate, for example, an electric bell and a relay.
◆ A current-carrying wire experiences a force when placed in a magnetic field.
◆ The direction of the force on a current-carrying wire depends on the direction of the current and the magnetic field.
◆ A d.c. electric motor requires a commutator.
◆ A commutator is an automatic switch which changes the direction of the current through the rotating coil every half revolution.
◆ Brushes are required in an electric motor to provide a good electrical connection between the commutator and the wires connected to the battery.
◆ Commercial motors use several coils of hundreds of turns of wire to create a greater turning effect. Each coil is connected to its own pair of commutator segments, which gives a greater turning effect and smoother running.
◆ Commercial motors have electromagnets instead of a permanent magnet. Electromagnets are more compact and produce a stronger magnetic field than a permanent magnet.

I Name three electrical appliances that use electric motors.

2 A relay is shown in the figure. When switch S is closed, lamp X switches on. Describe how the relay allows the lamp to light.

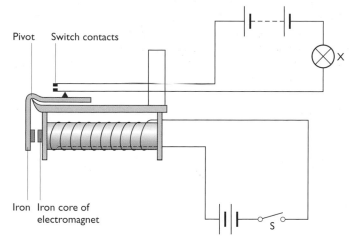

Figure 2.6Q2

3 The figure shows the main parts of an electric motor.

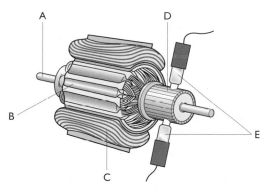

Figure 2.6Q3

Which letter represents the:

(a) brushes?
(b) commutator?
(c) field coils?
(d) axle?
(e) rotating coils?

4 The figure shows a simple electric motor. The coil has a clockwise rotation when connected in the circuit as shown.

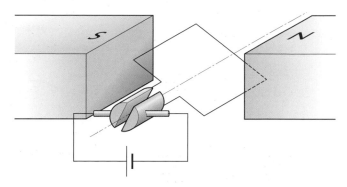

Figure 2.6Q4

(a) What will be the effect of making the following changes to the electric motor:
   (i) increasing the current through the coil?
   (ii) reversing the direction of the current through the coil?
   (iii) reversing the poles of the magnet?
(b) Explain the function of the commutator.

5 Give three reasons why carbon is a good material to use for the brushes of an electric motor.

1 The rating plate for a vacuum cleaner is shown below.

Model FJB19599

230 V a.c.
1380 W
50 Hz

Figure EQ2.1A Rating plate

a) A suitable flex has to be connected to the vacuum cleaner.
(i) Which one of the following flexes, W, X, Y and Z, is the most suitable for connection to the vacuum cleaner?

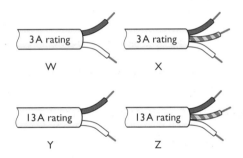

3 A rating   3 A rating
W   X

13 A rating   13 A rating
Y   Z

Figure EQ2.1B

(ii) Give two reasons for your choice in **a) (i)**.
b) Calculate the resistance of the vacuum cleaner when operating at its stated rating.

2 The diagram shows three electrical appliances connected to the mains supply in an unsafe manner.

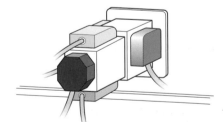

Figure EQ2.2

Explain why connecting appliances in this way could pose an electrical hazard.

3 A hi-fi player has a power rating of 15 W and operates from the a.c. mains supply.

a) State the declared a.c. value of mains voltage.
b) When the hi-fi is operating, calculate the current from the mains supply.
c) The hi-fi when switched on for 20 minutes. How much charge flows through the hi-fi in 20 minutes?
d) The hi-fi is connected to the a.c. mains supply and switched on. Describe the movement of charge in the hi-fi.

4 A student uses the circuit shown below to calculate the resistance of a resistor.

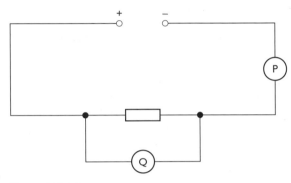

Figure EQ2.3

a) What does meter P measure?
b) The ammeter reading is 0.15 A and the voltmeter reading is 8.0 V. Calculate the resistance of the resistor.

5 A set of ten Christmas tree lamps is connected in series. The lamps do not light as the result of a lamp filament being broken. The broken filament cannot be seen.

The tree lamps are disconnected from the mains supply and a multimeter used to identify the faulty lamp. The multimeter is connected across each lamp in turn.

a) What quantity should the multimeter be set to measure — current or voltage or resistance?
b) How can the multimeter readings be used to identify the faulty lamp?
c) Would the ohmmeter indicate a short circuit or an open circuit when connected across the faulty lamp?

6 An electric cooker has two heating plates. Each heating plate is made up of two heating elements, each of resistance 100 Ω. The heating elements are connected in the circuits as shown below.

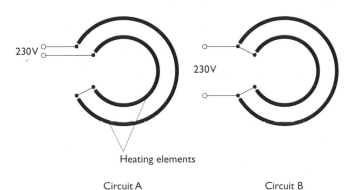

Heating elements

Circuit A   Circuit B

Figure EQ2.4

a) Which circuit, A or B, shows the heating elements connected in series?
b) Calculate the resistance of: (i) circuit A, (ii) circuit B.
c) Both circuits are connected to the 230 V mains supply. Calculate the current in: (i) circuit A, (ii) circuit B.
d) Which circuit, A or B, would heat a pot of soup in the shortest time? You must justify your answer.

7 a) Three resistors and an ammeter are connected in series to a 10 V supply as shown below.

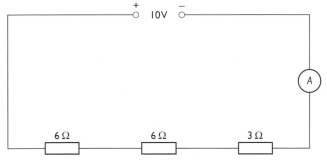

Figure EQ2.5

(i) Calculate the combined resistance of the three resistors in series.
(ii) The reading on the ammeter is 0.75 A. Show, by calculation, if the ammeter reading is correct.

b) The same three resistors are now connected in parallel to the 10 V supply as shown below.

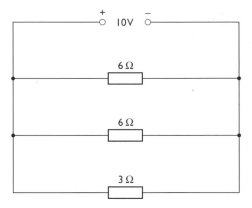

Figure EQ2.6

(i) Calculate the combined resistance of the three resistors in parellel.
(ii) Calculate the current drawn from the supply.

8 The element of an electric fire is operated from the 230 V mains supply. The mains flex connected to the element has three wires. The flex and part of the layout of the fire are shown below.

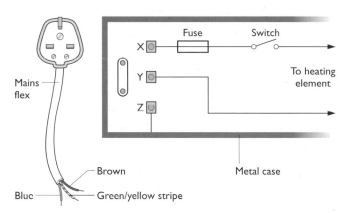

Figure EQ2.7

a) Which wire in the flex should be connected to
  (i) terminal X?
  (ii) terminal Y?
  (iii) terminal Z?
b) The element of the fire is rated at 1250 W.
  (i) When operating, what is the current in the element of the fire?
  (ii) The cost of 1 kWh of electrical energy costs 6.5 p. Calculate the cost of using the fire for 4 hours.

9 **a)** The main parts of an electric motor are shown below.

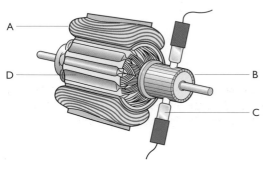

Figure EQ2.8

Some parts of the motor are listed below:

Brushes, commutator, field coils (magnet), rotating coil.

Use this list to identify parts A, B, C and D on the diagram.

b) The figure shown below shows a simple electric motor with a coil free to spin about shaft XY.

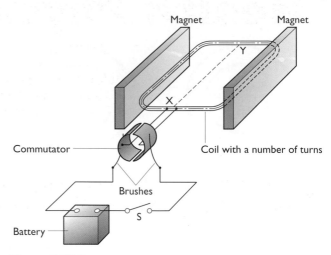

Figure EQ2.9

When switch S is closed, the coil rotates in a clockwise direction.
State **two** changes that would allow the coil to rotate in an anti-clockwise direction.

CHAPTER THREE
Health Physics

This chapter describes and explains how physics is used in the detection and treatment of different health problems. A large amount of work in health physics is concerned with looking inside the body without surgery.

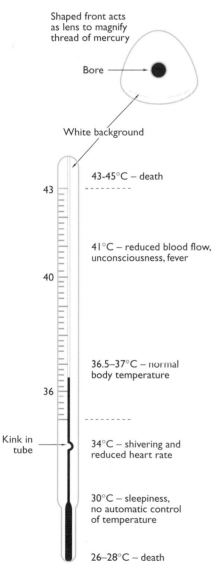

Shaped front acts as lens to magnify thread of mercury

Bore

White background

43-45°C – death

43

41°C – reduced blood flow, unconsciousness, fever

40

36.5–37°C – normal body temperature

36

Kink in tube

34°C – shivering and reduced heart rate

30°C – sleepiness, no automatic control of temperature

26–28°C – death

Figure 3.1 Clinical thermometer.

At the end of this section you should be able to:

1 State that a thermometer requires some measurable physical property that changes with temperature.
2 Describe the operation of a liquid in glass thermometer.
3 Describe the main differences between a clinical and ordinary thermometer.
4 Describe how body temperature is measured using a clinical thermometer.
5 Explain the significance of body temperature in diagnosis of illness.

Body temperature

Normal body temperature is 37°C. This is an average temperature and can vary by 0.5°C either way. It can rise to a peak during the day with maximum activity and falls when we are asleep. Most people register a temperature somewhere between 36°C and 37°C most of the time. A change of 3°C above that temperature can be dangerous and requires medical attention. One of the easiest methods of checking for many illnesses is to take your body temperature. Any changes from the normal temperature is an indication of possible concern.

Thermometers

Any substance which changes its properties with a change in temperature can be used as a thermometer. Most thermometers use the expansion of either a liquid or a solid. This property can be measured and so can be used to indicate temperature. There are a number of different kinds of thermometer.

Liquid-in-glass thermometer

This uses the expansion of a liquid to measure temperatures. The greater the temperature the greater the expansion of the liquid (figure 3.1). In the figure, as the mercury gets warmer, it expands along the glass tube, where there is a scale of numbers marked in °C (degrees Celsius).

The thermometer should be able to show small changes in the temperature by producing a movement of mercury which can be easily seen.

This means that the thermometer should be sensitive. A thermometer which is sensitive will have a narrow tube or bore and a large bulb.

To make the thermometer able to respond to rapid changes in temperature, it should have a bulb made of thin glass so that the heat can get through easily.

An ordinary thermometer cannot accurately measure body temperature since:

◆ The range is too large to measure small temperature changes of the human body.
◆ When the thermometer is removed from your mouth the liquid in the tube starts to fall, changing the reading.

Clinical thermometer

In the clinical thermometer the scale is from 35–42°C. The bulb is thin at the front to allow heat to flow quickly into it, but it is thick at the back to make it difficult to break.

It is designed to indicate and continue indicating the maximum temperature of the body. It works on the principle that a liquid – usually mercury – expands when it is heated. The liquid expands more than the glass so that the column of mercury moves up the tube. The narrow tube gives a very fine mercury thread, which expands quickly over a large length for a small change in temperature, making the thermometer very sensitive.

To find a patient's temperature, the thermometer bulb is placed under the tongue or arm and left for a minute or more. As the mercury expands, it forces its way past the bend or kink in the tube and eventually stops rising. The thermometer is then removed from the patient's mouth.

The mercury immediately starts to contract but, because there is a narrow bend or kink in the tube, the mercury thread breaks leaving a short thread of mercury to indicate the maximum temperature recorded.

To reset a clinical thermometer it has to be shaken fairly vigorously – but with care!

Changes in your body temperature affect the body in different ways. This is shown in figure 3.1.

Sometimes clinical thermometers have a digital scale powered by a battery. A special sensor is used instead of the liquid in the glass bulb, (figure 3.2a).

Other thermometers use liquid crystals which change colour when the temperature changes. These thermometers can be placed on the head to give the body temperature (figure 3.3).

Another type of thermometer uses the heat radiated from the body. A small sensor measures the heat from the ear and allows a baby's temperature to be easily found.

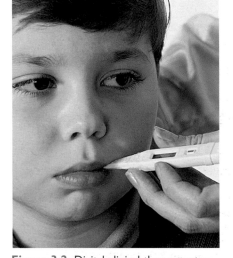

Figure 3.2 Digital clinical thermometer.

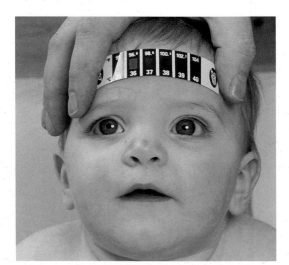

Figure 3.3 Liquid crystal forehead thermometer.

Body temperature and changes

Under normal conditions the body maintains the central organs such as the heart, lungs, abdominal organs and brain, at a fairly constant temperature of about 37°C, known as the 'core' or simply 'body temperature'. A lowering of body temperature is known as **hypothermia** and can happen by exposure to cold and damp conditions.

A more gradual deterioration in body function is observed during exposure to high temperatures caused, for example, by hot surroundings, heat therapy, fever or vigorous exercise. The condition, known as **hyperthermia**, is shown by blood vessels dilating, increased heart rate and cardiac output, and reduced blood flow to the brain which may result in unconsciousness. At about 41°C, the central nervous system starts to deteriorate, convulsions begin, and finally death occurs between 43°C and 45°C.

During a fever the high temperature speeds up the body's fight against infection.

Extreme body temperatures

Fascinating Physics

If you travel abroad you may suffer from malaria. This means that you will sweat and may become unconscious. Your temperature could reach 39°C. With suitable drugs you can recover quickly.

Hyperthermia can occur by extreme exertion in very warm conditions as described below. In very warm weather the human body cannot cope with overheating. This happened during a fun race in Australia in 1988 when a runner's body temperature rose to 42.8°C. The passage from the Sydney Morning Herald describes what happened to this athlete.

How a fun run meant meltdown for Mark Dorrity's body

By Sue-Ellen O'Grady

On February 27 this year, Mark Dorrity set off on what he expected to be an easy 8-kilometre fun run in Wagga, southern NSW. But near the finishing line, the fit 28-year-old collapsed, his body destroyed.

In less than an hour, his thigh muscles had overheated, liquefied and died. One leg has since had to be amputed at the buttock, because of gangrene.

Before Mark collapsed, his kidneys failed because the dying muscles had released toxic proteins into his blood which thickened to a molasses-like consistency. Every organ in his body was affected.

He suffered brain damage. His lungs could not function unaided. His buttock and hamstring muscles also liquefied, but not as severely as his thigh muscles.

Mark's heart stopped at least once. When it started again, it hammered away at 150 beats a minute, compared to its normal 70. He was on a dialysis machine for eight weeks, and in a coma for three months. When he regained consciousness, he could not walk or talk.

Even now, five months later, Mark cannot turn over or get out of bed unaided. He faces months of intensive rehabilitation.

The devastating damage to Mark Dorrity's body was caused by heat exhaustion and dehydration resulting in a condition known as rhabdomyolysis, the extreme result of what every runner and athlete knows as muscle fatigue.

The director of research at the Sports Medicine Institute, Dr Tony Miller, says the condition usually affects runners taking on more than they are used to in training.

"In Mark's case it was caused by his body severely overheating – to 42.8 degrees. At the same time, he was extremely dehydrated. When someone has a temperature that high, they are delirious. They ignore the body's warnings to stop."

Mark Dorrity was no weekend jogger. When he graduated from the University of NSW with an honours degree in Science in 1974, he won a Blue for athletics. He moved to Melbourne to work as a wool exporter, and ran four kilometres through the Botanical Gardens every day. As well, he swam a kilometre three times a week. He had minimal body fat.

He travelled to Wagga in February with a group of friends, all planning to compete in a local event. When the temperature that day rose to 42 degrees, the locals cancelled their run.

But Mark and his friends, deceived by the dry heat, decided to hold their own race.

"It just didn't feel that hot," he remembers. "So we ran off."

He drank several glasses of water before beginning to run, but none during the race. That, say doctors, proved to be his near-fatal mistake.

When Mark collapsed, he was leading the race by a kilometre. Friends driving alongside rushed him to the local hospital, where he was packed in ice to lower his body temperature. He remained there for two weeks before being moved to St Vincent's for dialysis treatment.

Mark recalls nothing of this. "I don't recall collapsing. I remember waking up twice at Wagga Hospital, and then waking here."

"How do I feel? I'm very lucky to be alive. I know that. I'm a medical miracle. And it's a warning to other runners to be extremely careful.'

The Sydney Morning Herald 3.8.1988

In a similar way, hypothermia can happen because of extreme cold weather and insufficient clothing. This happened to a child, Erika Nordby, in Canada.

Frozen alive

By John Harlow and John Elliot

Tiny footprints in the snow still reveal where 12-month-old Erika Nordby, dressed only in a nappy and a pink T-shirt, wandered outside before being frozen solid in the small hours of the morning eight days ago.

By the time Leyla Nordby, the infant's 26-year-old mother, woke up at 3 a.m. and realised her tot had slipped away from her side, Erika's heart had stopped, her last breath hanging in the freezing air, at –30° C.

It took the panicking mother, staying with a former school friend in Edmonton, Canada, nearly an hour to realise the back door to the house was ajar. Screaming and grabbing a torch, Leyla followed the footprints through an open garden gate, across the yard and into a secluded corner, 30 ft from the back door. Her daughter lay face down in the foot-high snow.

The child's eyelids were frosted with snowflakes, her toes were frozen together and her mouth was sealed shut by an icy bubble of spittle on her lips. Torchlight showed that her arms and legs were scalded bright red with signs of frostbite.

The ambulance arrived within 10 minutes and paramedic Krista Rempel checked for vital signs: there was no pulse and Erika's body temperature had dropped to 16° C – less than half the normal 37° C. It was immediately apparent to the veteran medic that blood was no longer flowing through Erika's heart, pooling instead in her face and belly on which she had fallen.

Such occurrences are not uncommon. About 20 people a year succumb to hypothermia in the winter wastes around Edmonton, a death that

North American writer Jack London described a century ago as 'the most gentle doorway into the night'. If it had been anyone but a toddler, the paramedics might have been ready to declare the cold body officially dead.

This, however, was a 25 lb, previously healthy toddler who had been born at the University of Edmonton Hospital on February 16 last year. By a lucky coincidence, too, Rempel had seven years previously helped to revive another little Canadian girl who was found frozen solid 'like a lollipop'.

Still haunted by the image of adrenaline needles breaking on that child's frozen skin, Rempel cracked Erika's lips apart to push a breathing tube into her throat and wrapped her in a 'bear-hug' blanket to warm her up. She then put the university's Stollery children's health centre on full alert.

They knew what to do. Shortly after 4 a.m., Erika was being attached to a heart-lung machine in preparation for a warming blood transfusion when the unexpected happened. Her heart started beating again, suddenly and spontaneously. She was fighting back to life.

Dr Alf Conradi, director of medicine at the Stollery, said that Erika's heart had been stopped for perhaps two hours by the time she reached the hospital. 'But we were not ready to give up on her. We knew that, with hypothermia victims, there is always a slim chance of revival by shocking the heart back into action, but this time the heart did it by itself. I do not usually use words like miracle but it is appropriate in this case,' he said.

His colleague, Dr Allan De Caen, part of the emergency team that helped save Erika, said that a little body can often stand a better chance of surviving the big chill than larger ones: 'Erika's organs and blood would have frozen much more quickly than an adult's organs or blood, which meant that there was less damage. As she went into a coma, the amount of oxygen that her brain needed fell sharply. It may have been more like hibernation than death, something we do not fully understand, although by every textbook definition she was dead.'

This weekend Erika was in a private ward at the Stollery, playing with dozens of stuffed toys sent by well-wishers from around the world. Her only complaint was about not being allowed to watch Barney the Dinosaur, her favourite television programme.

Her hands and feet are wrapped in bandages that are changed at least four times a day, the only time she is in severe pain, and her cheeks are still rosy, marked by the blood that pooled in her face as she lay in the snow.

Doctors say she is recovering quickly. Her hands are healing, although arm and leg skin grafts are under consideration and plastic surgery may still be needed on her toes.

The Sunday Times 30/05/01

Hypothermia and fever are two examples of your body temperature falling very low or rising very high. But even small changes in temperature can have a big effect on your body. If your body temperature falls by as little as one degree you begin to shiver. This helps to heat you up. The blood vessels near the surface of the skin will retreat from the skin and the extremities such as the feet and the hands – which is where most frostbite occurs. The reverse occurs when the body tries to get rid of heat when the blood vessels will increase the blood flow to the skin.

Section 3.1 Summary

- ◆ A thermometer requires some measurable physical property that changes with temperature, a typical property being the expansion of a liquid.
- ◆ A liquid in glass thermometer has a thin bore within a glass tube which contains a liquid in the bulb.
- ◆ A clinical thermometer has narrow range of temperature scale compared to a normal thermometer, and a small kink in the tube to prevent the liquid falling back when it is removed from the patient to read the temperature.
- ◆ Body temperature is measured by placing the thermometer in the mouth and waiting for a suitable time until it has reached the body temperature.
- ◆ The significance of measuring body temperature is when it changes from 37°C and can indicate fever or shivering.

End of Section Questions

1 Describe two important features of a liquid in glass thermometer, which allow you to take a body temperature.

2 Explain why the measurement of body temperature gives important information to a doctor.

3 During a mountain climb Arthur's body temperature is measured as 34°C. Explain whether this value should give cause for concern .

At the end of this section you should be able to:

1 State that a solid, a liquid or a gas is required for the transmission of sound.
2 Explain the basic principles of a stethoscope as a 'hearing aid'.
3 Give one example of the use of ultrasound in medicine, for example images of an unborn baby.
4 State that high frequency vibrations beyond the range of human hearing are called ultrasounds.
5 Give two examples of noise pollution.
6 Give examples of sound levels in the range 0–120 dB.
7 State that excessive noise can damage hearing.
8 Explain one use of ultrasound in medicine.

Listening using a stethoscope

The stethoscope is a hearing aid which allows a doctor or nurse to listen to sounds made within the body. It is an old and basic instrument which can give some vital information very quickly. It is most often used to listen to the heart and the lungs. These sounds can be useful in the diagnosis of various diseases.

The main parts of a modern stethoscope are shown in figure 3.4.

◆ The chestpiece has two 'bells', one open and the other closed by a thin diaphragm (a semi-rigid disc). A valve can be turned to change from the open to the closed bell.
◆ The open bell is used to listen to heart sounds.
◆ The closed bell is used to listen to sounds which have a higher frequency than heart sounds, such as those from the lungs.
◆ Sounds picked up by the open bell or the diaphragm are transmitted to the earpieces through the air in the tubing. The eardrum of the listener is also a pressure sensitive diaphragm. To create sufficient pressure change at the ear for a given movement of the diaphragm it is important to have a bell with as small a volume as possible. The volume of the tubes should also be small and this requires short tubes with a small diameter. But we also need to have very little loss of sound due to friction and this requires large tube diameters!
◆ The earpieces have to be a good fit with the ears to avoid sound losses and to prevent background sounds from interfering with those from the heart and the lungs.

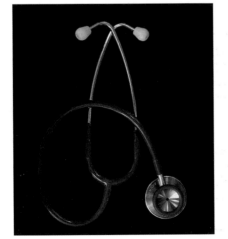

Figure 3.4 A modern stethoscope.

Ultrasound – sounds beyond your hearing

Young people can hear sounds with frequencies from 20 to 20 000 Hz. This is the normal range of frequencies but as we grow older the upper range of frequencies decreases. Frequencies above 20 000 Hz are called ultrasound or ultrasonic vibrations. When these higher frequency ultrasonic waves are sent out by a transmitter and hit an object, some of the waves will pass through the object while some will be reflected. To find their way about, bats, whales and dolphins use the echoes from the reflections. Submarines and fishing boats use a similar system called SONAR (SOund Navigation And Ranging). This allows them to determine the depth of water or the location of different objects. As the signals hit the sea floor the reflected signal will come back sooner and the depth can be calculated (figure 3.5)

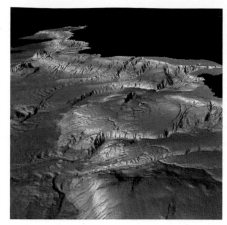

Figure 3.5 A sonar image of the sea floor off the coast of California, USA

In medicine the frequencies used are between 1 and 20 MHz (M = mega = 1 million). The waves are sent out from a transducer. This is a simple hand-held device which changes one form of energy into another form. The transducer also acts as a receiver which can pick up the reflected waves. In this case electrical energy is changed to sound energy in the transmitting part and the opposite when acting as a receiver.

Very high frequencies of between 1 and 5 megahertz (MHz) are used in medicine. The speed of sound in soft biological tissue is about 1500 m/s and the distances being measured are of the order of 0.15 m, so the echo time is around 200 microseconds (µs) (200×10^{-6} s = 0.0002 s). Check this calculation yourself.

Ultrasonic pulses are sent into the body from a transmitter placed in good contact with the skin. Reflections then come back from any boundary between materials of different sound properties. The more ultrasound that is reflected from the surface of the body, the less detail that can be seen of structures in the body. Reflections happen where there is a large change in the structure, for example from bone to muscle or from bone to other soft tissue. Some waves must be transmitted into the body so good contact between the transmitter and the skin (body surface) is very important. If most of the ultrasound is reflected at the boundary between the air and the skin, little of the ultrasound would enter the body. To make good contact between the transmitter and the skin, and so allow most of the ultrasound into the body, a gel is smeared on the patient's skin. Once inside the body, the pattern of reflected ultrasound can be used to build up a picture of the object inside the body (figure 3.6). The display on the screen in figure 3.6 is formed by the distances and intensities of the echoes. This forms a two dimensional image of the object.

The greater the frequency the smaller the wavelength and the greater the detail that should be seen. However, the greater the frequency the more the waves are absorbed by the material and the amount of energy transferred as heat increases. Different frequencies of ultrasound are used for different organs in the body. At the chosen frequency the maximum distance travelled will be about 200 wavelengths. This allows the greatest amount of detail to be seen.

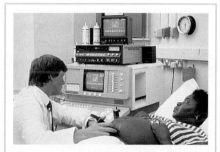

Figure 3.6 Use of ultrasound.

A typical scan is shown in figure 3.7 which checks on the progress of a baby in the womb. By measuring the diameter of the head we can check the age of the baby. Another use of ultrasound scanning is to check the functioning of the valves in the heart. The thickness of the eye lens can also be measured, which is often used in eye examinations (figure 3. 8).

The main advantages of ultrasound are:
◆ No surgery is needed to see inside the body.
◆ Since ultrasound operates at low power levels no harmful effects on the body have been found and it is safer than X-rays. This means that it can be used safely for checks on a foetus.

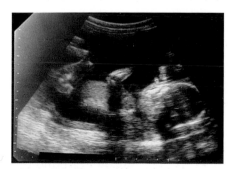

Figure 3.7 Scan of a foetus.

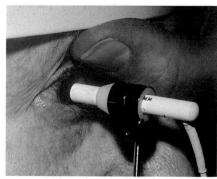

Figure 3.8 Scanning the thickness of the eye lens.

Doppler ultrasound

Ultrasound has developed in different ways in recent years. One main technique is called Doppler ultrasound. This uses a technique first noticed by Christian Doppler. As a car siren approaches you may notice that the pitch or frequency of the sound changes and alters again as the car drives away from you. As a moving object reflects the ultrasound waves it changes the frequency of the reflected waves. This is called the 'Doppler effect'. The change in frequency increases as the speed of the object changes. (You will also meet this technique in Chapter 5 in the measurement of the speed of cars.)

This change in frequency is used by ultrasound pulses to measure the speed of blood flow through the heart (figure 3.9). The direction of blood flow can be indicated as different colours. Red indicates the blood flowing towards the transducer and blue flowing away. The different shades of blue and red indicate the different speeds. The method is used for final examinations for defects in heart valves before heart surgery when X- rays would be dangerous.

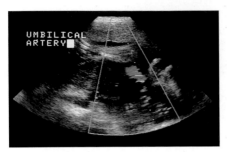

Figure 3.9 A doppler ultrasound of a foetus showing bloodflow through the umbilical cord.

Losing stones – in minutes!

Ultrasound has been used to treat patients with kidney stones. These are particles of material which may form in the kidneys. The stones block the flow of waste material and can cause excruciating pain. Instead of surgery the stones can be broken up by a lithotripter (figure 3.10). This uses several ultrasound beams focused on the stones so that shock waves are produced at a common focus. The stones are shattered and not the surrounding tissue. The patient has small pads attached to him or her through which the pulses are directed. When the stone is broken into small pieces they are then passed out in the normal waste from the body.

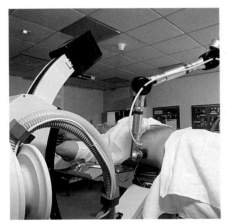

Figure 3.10 Lithotriper.

Measuring sound level

The human ear is a sensitive detector of sound and can be damaged by very loud sounds. The loudness of sound is measured in bels named after Alexander Graham Bell. Since the bel is a large unit we use **decibels (dB)**, which is one tenth of a bel. These levels are measured using a sound level meter (figure 3.11).

The decibel scale is not like a normal scale on a ruler. If one sound is 10 times more powerful than another then it is said that it is 10 dB greater. If it is 100 times greater then it is 20 dB greater. The power produced by sound is actually very small. At a typical football or rugby game the sound produced by shouting would only provide sufficient energy to heat a cup of coffee.

Noise pollution

Noise is unwanted sound. It may be sound from traffic, from a neighbour's TV or radio, from machinery at work and so on. Because it is unwanted, noise is a kind of pollution.

Typical noise levels are shown in the table 3.1.

The first danger level is from 85 to 90 dB. Everyone agrees that excessively loud noises (that is above 90 dB) can be unpleasant and some can cause damage to the hair cells in the inner ear. There is, however, disagreement about the damage caused by certain kinds of sounds – for example disco music.

In factories or noisy workplaces, near pneumatic drills and aircraft, or in heavy vehicles or tractors the noise level can be over 100 dB. This can cause permanent damage to the ears with a serious loss of hearing ability. A 'ringing' sound heard after exposure is a warning sign. Some people become irritable, short tempered and tired as a result of exposure to very loud noises. Ear protection – ear-plugs, ear-muffs or a helmet – can be used

Figure 3.11 A sound level meter.

Fascinating Physics

Source of sound	Sound level (dB)
Jet engine at 50 m	130
Disco, 1 m from loudspeaker	120
Pneumatic drill at 5 m	100
Heavy goods vehicle from pavement	90
Alarm clock 0.5 m from bedside	80
Telephone ringing at 2 m	70
Vacuum cleaner at 3 m	70
Normal conversation at 1 m	60
Boiling electric kettle at 2 m	50
Residential area at night	40
Quiet country lane	20
Silence (hearing threshold of humans)	0

Table 3.1 Typical noise levels.

Figure 3.12 Ear protectors being used.

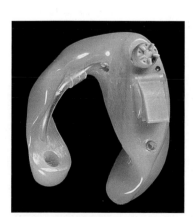
Figure 3.13 A hearing aid.

in many cases to reduce the level of noise heard (figure 3.12). These devices work by filtering out the sound by using a thick material to reduce the sound level. Noise coming through a window can be reduced by about 20 dB if double glazing is fitted.

The silencer of a motor car exhaust system is designed to trap the sound. In particular the high frequency sounds from the explosions in the engine are absorbed. In the USA this is called a muffler. A new silencer can reduce the sound level from 150 dB to 85 dB.

The small bones which transmit vibrations from the eardrum to the inner ear stiffen up and transmit the vibrations less effectively at sound levels above 80 dB. This, to some extent, protects the ear from loud sounds. However, the ear's reflex action takes time to operate and it is during this time, of less than 1 second, that the short peaks of sound can pass into the inner ear and cause damage. The noise level and how long it lasts will affect your hearing and the law limits exposure to noise to no more than 90 dB for eight hours and 93 dB for four hours, 96 dB for two hours, and so on.

Hearing loss

As we grow older we all experience some deafness. Indeed you may notice this in some people already since 1 in 10 people will be experiencing some difficulty. This will rise to 1 in 4 if you are over 65.

Hearing impairment varies from total deafness, usually caused by a defect at birth or by disease or accidental damage to the ear mechanism, to the slight impairment which often comes with age. Hearing aids can be used to correct some of these problems and restore almost normal hearing. The latest hearing aids amplify normal speech frequencies but filter out background noise. Hearing aids consist of a small microphone, amplifier, battery and earphone. (figure 3.13). For many hearing-impaired people their hearing loss is greater at certain frequencies.

About 1 in 5 of the adult population have hearing that is substantially impaired. This is defined as a decrease in intensity levels of 25 dB over the speech range of 500 to 2000 Hz. Six different frequencies are used to test for any hearing loss.

- A solid, a liquid or a gas is required for the transmission of sound.
- A stethoscope uses two bells to pick up different frequencies. The structure of the tubes increases the volume of the sounds.
- An example of the use of ultrasound in medicine is in the imaging of an unborn baby.
- High frequency vibrations beyond the range of human hearing, which is beyond 20 000 Hz are called ultrasounds.

- Ultrasound works by using the reflections from structures to detect soft tissue. If the ultrasound is not transmitted then the structure will not be seen clearly.

- Noise pollution is unwanted noise.
- Excessive noise can damage hearing.

End of Section Questions

1 What are the key parts of a stethoscope and what is the purpose of each part?

2 (a) What is meant by ultrasound ?
   (b) Give an example of its use in medicine.

3 Ultrasound is transmitted through soft tissue at a speed of 1460 m/s. The ultrasound takes 3 μs (microseconds) to echo from reflection from a bone. Calculate the distance from the tissue to the bone.

4 During a study class, the noise of a pneumatic drill is heard through an open window. Suggest a value for the noise level experienced a few metres from the drill. If this noise persists for a short time what might be the physical effect on the person's hearing?

At the end of this section you should be able to:

1 Describe the focusing of light on the retina of the eye.
2 State what is meant by refraction of light.
3 Draw diagrams to show the change of direction as light passes from air to glass and glass to air.
4 Describe the lens shapes of convex and concave.
5 Describe the effect of various lens shapes on the rays of light.
6 State that the image formed on the retina of the eye is upside down and laterally inverted.
7 Explain using a ray diagram how an inverted image can be formed on the retina.
8 Describe a simple experiment to find the focal length of a spherical convex lens.
9 State the meaning of long and short sight.
10 State that long and short sight can be corrected using lenses.
11 State that fibre optics can be used as a transmission system for 'cold light'.

12 Use correctly in context, the terms angle of incidence, angle of refraction and normal.
13 Explain using a ray diagram how the lens of the eye forms, on the retina, an image of an object (a) some distance from the eye and (b) close to the eye.
14 Carry out calculations on power/focal length to find either one given the other.
15 Explain the use of lenses to correct long and short sight.
16 Explain the use of fibre optics in the endoscope (fibrescope).

Refraction of Light

Many people need to wear spectacles or contact lenses and we will nearly all need to have some help with reading books like this as we grow older, but not just yet for you hopefully. Refraction is the term used when light passes into or out of a piece of glass. This effect can explain how some sight defects such as long and short sight occur and how they can be corrected.

In a given material (called a 'medium') light travels in a straight line. When the light travels from one material to another it may bend as it enters the new material. When light enters a new material its speed changes. This effect is called **refraction**.

Rays of light can travel through various objects. The paths of the rays passing through and leaving the objects can be drawn as shown in figure 3.14. The dotted line drawn at right angles to the surface is called the normal. All angles are measured against the normal.

Plane rectangular block

When the incident ray travels parallel to the normal, there is no change in direction.

When the incident ray is at an angle to a plane rectangular block, the ray coming out of the block is parallel to the incident ray. (figure 3.15). The angle in the block is called the angle of refraction and is less than the angle of incidence.

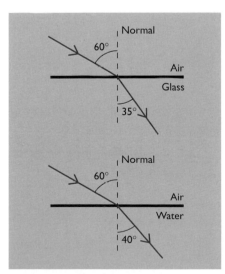

Figure 3.14 Light travels more slowly in glass than it does in water. So a light ray bends more when it goes into glass.

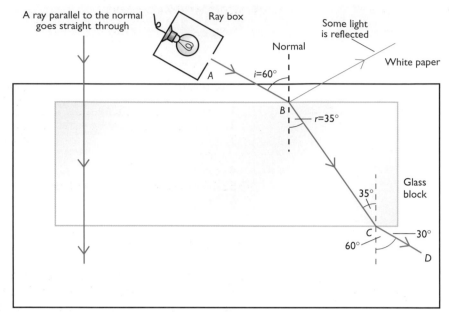

Figure 3.15 The path of a ray through a plane rectangular block.

Triangular prism

The ray bends towards the normal going into the prism and away from the normal coming out (figure 3.16).

Convex or converging lens

The middle ray goes straight on and the outer rays bend and meet on the middle line at a point called the focus (figure 3.17). If the lens is thick, the same effect occurs but the focus is nearer the lens (figure 3.18).

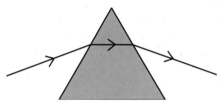

Figure 3.16 A ray travelling through a prism.

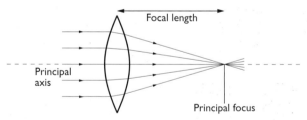

Figure 3.17 A thin lens is a weak lens; it has a longer focal length than a strong lens.

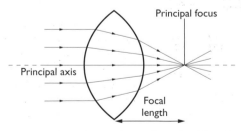

Figure 3.18 A thick lens is a strong lens; it has a short focal length.

Concave or diverging lens

The rays spread out (figure 3.19).

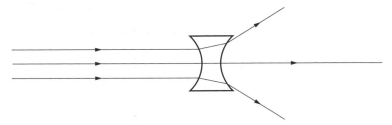

Figure 3.19 A concave or diverging lens.

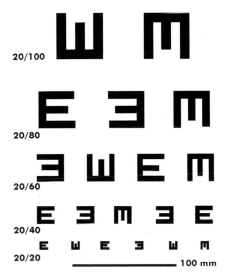

Figure 3.20 The human eyeball.

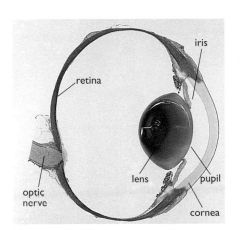

Figure 3.21

The eye

Our eyes tell us what's going on in the 'outside world'. They enable us to grasp, hit or touch objects 'out there' and, of course, to avoid being grasped, hit or touched by threatening objects. Our eyes tell us the shape, size and colour of objects. An outline of the eye is shown in figure 3.20. The different parts of the eye and their functions are:

♦ Light enters the front of the eye at the **cornea**. This is transparent and most of the refraction of light occurs here. The eye can detect changes from bright daylight to a dark night, which is a ratio of 10 million to one!

♦ The light enters a **lens**, which is a jelly-like substance, and more refraction takes place. The lens is held in place by fibres which act like muscles. These can change the shape of the lens from thick to thin.

♦ The light then passes through a gel-like substance which makes the light spread out.

♦ The light reaches the **retina** at the back of the eye. This has special cells to receive the light.

♦ The retina has about 100 million tiny nerve endings (cells) called **rods** and **cones**. The rods are sensitive to small amounts of light and are used in night vision. The cones give us colour vision and help us to see detailed sharp images. They detect red, green and blue. In poor light, the cones do not function and we tend to see objects in shades of grey.

♦ The area of sharpest vision is called the **fovea** or 'yellow spot'. It is packed with cones and therefore responds well to sharp, bright, coloured images.

♦ The **iris** controls the amount of light entering through the eye by the **pupil**, which can alter in the size of the opening.

Electrical signals pass along the nerve fibres to the brain. The part of the retina where the nerve fibres leave the retina contains no light-sensitive cells and is therefore a **blind spot** on the retina.

An eyesight chart to test vision is shown in Figure 3.21.

The amount of refraction which takes place at the cornea does not change. However, in order to focus on near and on distant objects, an adjustable lens is needed. This is provided by the eye's lens. This lens is held by muscle like fibres in the **ciliary body** which can change the shape of the lens from thick to thin. The lens is thin and can focus on distant objects. To view near objects the muscles change the lens shape to thick. When light enters the eye, the image formed on the retina is upside down. The brain learns to turn this image the right way up.

Focal length

Some lenses bend light more than others, due to their thickness and the amount of curving of the lens. One way to indicate the amount of refraction is to measure the focal length of the lens. A convex or converging lens can make rays of light come together to a point after they have passed through the lens. The point where the rays meet is called the **focus**. The position of the focus depends on where the rays come from. When the rays come from a distant object, which is so far away that the rays are parallel, the focus is closer to the lens. In this case, the focus is called the **principal focus**. The distance from the lens to the principal focus is called the **focal length** and is measured in metres.

Measuring the focal length of a lens

1 A converging lens is held near a window frame and the image of an object outside is brought to a focus on a piece of card.
2 Move the card until the image is sharp.
3 The distance between the lens and the card is measured with a ruler.
4 This distance is the focal length (figure 3.22).

Typical values for focal length of a school lens are in the range from 2 to 25 cm.

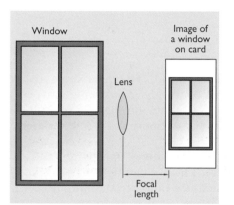

Figure 3.22 Measuring the focal length of a common lens.

Power of a lens

People who have severe eye defects may need stronger (more powerful) lenses to correct their eyesight than those who have only slight defects. (A more powerful lens is one which causes more refraction.) An optician must therefore have a range of lenses to suit different individuals' needs. The powers of these lenses could be indicated by giving their focal lengths – the most powerful lenses having the shortest focal lengths. Another way to indicate the amount of refraction caused by a lens is to calculate its power from the equation:

$$\text{Power} = \frac{1}{\text{focal length}}$$

Where the focal length is measured in metres and the power is given in dioptres (D).

◆ Converging (convex) lenses have positive powers (e.g. +10 D, +17 D).
◆ Diverging (concave) lenses have negative powers (e.g. –10 D, –17 D).

Example
A convex lens has a focal length of 25 cm. Find the power of the lens.
Solution

$$\text{Focal length} = 0.25 \, \text{m}$$

$$\text{Power} = \frac{1}{\text{focal length}}$$

$$= \frac{1}{0.25}$$

$$= 4 \, \text{D}$$

Example
A lens has power of –2.5 D. Calculate its focal length.

Solution
Power = –2.5 D, which tells us that this a concave lens since there is a negative sign.

$$\text{Power} = \frac{1}{\text{focal length}}$$

$$2.5 = \frac{1}{\text{focal length}}$$

$$\text{focal length} = \frac{1}{2.5}$$

$$= 0.4\,\text{m}$$

Long sight

A long-sighted person can see faraway objects clearly. Objects quite close to the eye appear blurred. The eye lens is bringing the rays to a focus beyond the retina. This may happen if the person's eyeball is shorter than normal from front to back. It may also be caused by ciliary muscles which cannot relax for the lens to be fat enough.

A converging (convex) lens corrects this fault since it will increase the bending of the light rays before they enter the eye lens. Light will then be focused on the retina and is thus seen clearly (figure 3.23).

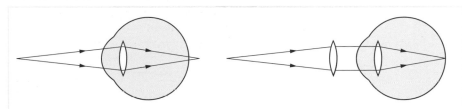

Figure 3.23 (a) A long-sighted eye cannot see near objects clearly, (b) a converging lens corrects long sight.

Short sight

A short-sighted person finds that distant objects are blurred but near objects are in focus. The eye lens is bringing light to a focus in front of the retina. This may happen due to the lens having a large curvature. The muscles cannot make the lens thin enough.

A diverging (concave) lens corrects this fault, since it will spread the light out more, before it enters the eye lens and so light will be focused on the retina (figure 3.24).

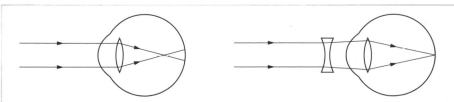

Figure 3.24 (a) A short-sighted eye cannot see distant objects, (b) a diverging lens corrects short sight.

Spectacle lenses and contact lenses

The first spectacle lenses were developed around about 1270 but there is a legend that the Emperor Nero had an emerald cut as a lens to enable him to see more clearly.

Contact lenses were first suggested by Leonardo da Vinci in 1581 who noticed that he could see more clearly when he opened his eye under a bowl of water. Different lenses have been developed which allow oxygen to pass into the eye which helps to prevent some eye diseases (figure 3.25). It is now possible to correct short sight by using contact lenses to remould the cornea. This involves a scan of the eye to determine the exact curvature. A reverse contact lens is used which is placed in the eye at night. The lens presses on the cornea and gradually reshapes it. It is not permanent and the lenses must be placed in every night. They are useful for sportspeople, fire workers or those who find conventional lenses irritating. There are no surgical procedures and the process is reversible. This technique is called orthokeratology.

Figure 3.25 Contact lenses.

Astigmatism

If a person suffers from astigmatism then the person has difficulty focusing light entering the eye in different planes. This is caused by a distorted cornea (figure 3.26). A cylindrical lens is needed for correction.

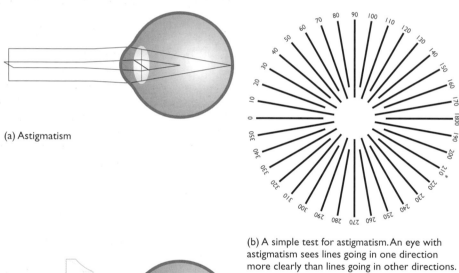

(a) Astigmatism

(b) A simple test for astigmatism. An eye with astigmatism sees lines going in one direction more clearly than lines going in other directions.

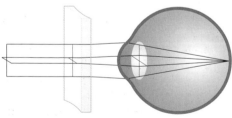

(c) Correction of astigmatism using a cylindrical lens

Figure 3.26 (a) Astigmatism, (b) a simple test for astigmatism: an eye with astigmatism sees lines going in one direction more clearly than lines going in other directions, (c) correction of astigmatism using a cylindrical lens.

A typical prescription for correcting eye problems is shown in table 3.2.

Right Eye						Sph	Cyl	Axis	Prism	Base
Sph	Cyl	Axis	Prism	Base						
−1.2	−0.7	90			Distance	−1.2			1Δ	1 N
+1.4	−0.7	90			Near	+1.5				

Table 3.2 Prescription for a short-sighted person.

This means that for the right eye:
◆ The term Sph stands for spherical and the negative indicates that this short-sighted person needs a lens of power 1.2 D.
◆ The correction of +1.4 D means that a converging lens is needed for near vision.
◆ The Cyl values correct astigmatism and the Axis value indicates that the axis of this lens should be at 90° to the horizontal.
◆ The Prism and Base values tell us that a prism is needed for both eyes to see in the same direction.

Seeing round the bend with a cold light source

Using fibre optics in medicine

Optical fibres are about the thickness of a human hair. Each fibre consists is a thin piece of glass coated with a thin layer or cladding of another glass. This cladding prevents the light, which enters the end of the fibre, from escaping or passing through the sides to another fibre in the bundle (see chapter 1). Fibres can be put together so that each fibre lies in the same direction as other fibres.

The fibres are called **coherent** and an image can be transmitted down these fibres. Those fibres which are arranged in a random way are called **incoherent**.

Optical fibres only allow light to be sent down them. These fibres in medicine are used in devices called endoscopes or bronchoscopes.

The key elements of these devices are:
◆ An incoherent bundle of fibres to send light down to the internal organs by total internal reflection.
◆ Light reflected from the organs to send an image up a coherent bundle of fibres.
◆ A channel to clean the lens (figure 3.27).

The heat from the lamp does not pass down the fibres. This means that the other end of the guide is cold (called a 'cold light source'). This is one of the advantages of the endoscope.

Endoscopes have a bending section near the tip so the observer can direct the instrument during insertion.

In recent operations the endoscope is used for keyhole surgery. This involves making a small insertion in the body to insert the endoscope and small instruments to perform the intricate surgery. The advantage is that the small incision means a faster recovery.

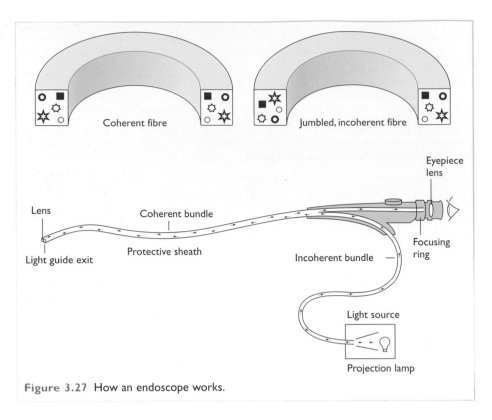

Coherent fibre

Jumbled, incoherent fibre

Eyepiece lens

Lens

Coherent bundle

Focusing ring

Light guide exit

Protective sheath

Incoherent bundle

Light source

Projection lamp

Figure 3.27 How an endoscope works.

Endoscopes can be used to view a tumour which might occur in the lungs (figure 3.28). An X-ray can show possible areas of concern but the use of a bronchoscope will show up the tumour.

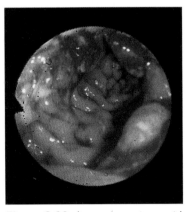

Figure 3.28 A tumour as seen with an endoscope.

Section 3.3 Summary

◆ Light is focused by the cornea and the lens on the retina of the eye.
◆ Refraction of light occurs when light travels from one substance to another.
◆ When light travels from a less dense to a more dense substance the ray bends towards the normal in the more dense substance.
◆ A convex lens brings parallel rays to a focus and a concave lens spreads the rays out.
◆ The image formed on the retina of the eye is upside down and reversed.
◆ Long sight is when distant objects are seen clearly but near objects are blurred. To correct this defect a convex lens is needed.
◆ Short sight is when distant objects are blurred but near objects are clear. A concave lens will correct this.
◆ Fibre optics can be used as a transmission system for 'cold light'.
◆ Power of a lens $= \dfrac{1}{\text{focal length in m}}$ and is measured in dioptres.

1 What is the function of the retina in the eye?

2 Describe how the image of a tree is seen on the retina compared to the original tree.

3 A ray of light enters a glass block as shown below. Copy and complete the diagram showing the passage of the ray through the glass and its emergence out of the other side.

Figure 3.2Q3

4 A convex lens has a focal length of 10 cm.

   (a) Explain what is meant by the term focal length.
   (b) Describe how you could measure the focal length of this lens.

5 Teri can see distant objects clearly but near objects are blurred.

   (a) What defect is she suffering from?
   (b) What kind of lens is needed to correct this problem.

6 A convex lens has a focal length of 12.5 cm. Calculate the power of this lens.

7 Keyhole surgery uses a fibre optics system to view inside the body. What is the advantage of using fibre optics rather than a conventional light source?

At the end of this section you should be able to:

1 Describe how the laser is used in one application of medicine.
2 Describe one use of X-rays in medicine.
3 State that photographic film can be used to detect X-rays.
4 Describe the use of ultraviolet and infrared in medicine.
5 State that excessive exposure to ultraviolet radiation may produce skin cancer.
6 Describe the advantage of computerised tomography.

Electromagnetic spectrum

The light that we can detect with our eyes is only a small part of all the wavelengths that exist. The range is called the electromagnetic spectrum. The visible spectrum ranges from 400 to 700 nm (a nm is a nanometre = 1/1 000 000 000 m). The complete spectrum is shown in figure 3.29. The explanation and detection of the different radiations is explained in Chapter 7.

All parts of the spectrum travel through space at a speed of 3×10^8 m/s (300 000 000 m/s). Each member of the electromagnetic spectrum has a different wavelength and frequency.

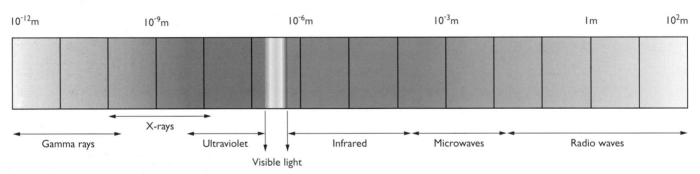

10^{-12}m 10^{-9}m 10^{-6}m 10^{-3}m   1m 10^2m

X-rays

Gamma rays   Ultraviolet   Infrared   Microwaves   Radio waves

Visible light

Figure 3.29 The electromagnetic spectrum.

Infrared radiation in medicine

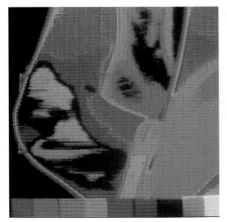

Figure 3.30 Thermogram of an arthritic joint.

All hot objects give off invisible 'heat rays' called **infrared radiation**. Special infrared cameras can be used to take colour photographs called thermograms using this radiation instead of light. The infrared radiation allows us to measure small temperature changes inside tissue without using surgery. In medicine, thermograms of a patient's body show areas of different temperature. Doctors have found that malignant tumours are warmer than healthy tissue and show up clearly on thermograms. If people suffer from arthritis then the affected joint will show as a different temperature from the normal joint (figure 3.30).

Figure 3.31 Heat-seeking camera mounted on the front of a rescue helicopter.

Another use of infrared radiation is the heat-seeking cameras used to detect people who may be trapped in buildings. (figure 3.31).

Infrared radiation is used in a different way by physiotherapists, who use it to treat people who have suffered a muscle injury. They use this radiation to penetrate the skin and heat muscles and tissues. Heat causes healing to occur more quickly.

Typical wavelengths of infrared radiation are from 700 to 1500 nm.

Ultraviolet radiation

Ultraviolet is another type of invisible radiation. The wavelength of ultraviolet rays is in the range from 200 to 400 nm. This is shorter than the wavelength of visible light in the range from 200 to 400 nm. Although it forms only 3 per cent of the solar light, it is the most active part of the spectrum in terms of damage by light. There are three types of UV radiation:

◆ UVA from 320 to 400 nm in wavelength;
◆ UVB from 290 to 320 nm in wavelength;
◆ UVC less than 290 nm.

All UVC and part of UVB is absorbed by the Earth's ozone layer. UVB is absorbed by glass and many types of plastic. UVB is the part of the spectrum that causes sunburn and is more damaging than UVA. We need UVA for healthy growth and to make vitamin D.

UVA is not screened by glass and until recently sunscreens were not effective against this radiation. UVA makes up more than 90 per cent of the UV radiation that reaches the Earth's surface. Excessive exposure to UVA and UVB can cause skin cancers called melanomas. There is also a risk from long exposure under sun beds, and in this case care should be taken to cover the eyes since they are especially sensitive to this damage. Too much ultraviolet light on the skin produces sunburn and can cause the skin to turn red and be very painful. In addition there is evidence that the skin ages more prematurely and that ultraviolet light can cause cataracts. Suntan lotions absorb some of the ultraviolet rays which cause the burning, but they allow the lower frequency rays to reach the skin and to produce a tan. The tan is due to a pigment called melanin being produced.

In the last few years the greatest rise in cancers in the UK has been as skin cancers. Your skin can 'remember' the last amount of exposure to the sun that it received and over a period of time this may lead to skin cancer. You may think the risk is not very great but in the USA out of a population of 200 million people, about 1 million people will suffer from this disease. It is more common in people with a light coloured skin who have spent a lot of time in the sunlight. You will receive more sunlight if:

◆ you are nearer the equator;
◆ you are at a higher altitude from sea level;
◆ you are sunbathing round about midday.

You can reduce the risk by:
◆ Limiting your time in the sun.
◆ Using a sunscreen lotion with a factor of 15 or greater. The number indicates the length of time that you can stay in the sun.
◆ Checking the UV index given in weather forecasts (table 3.3).

UV Index number	Exposure level
0 to 2	Minimal
3 to 4	Low
5 to 6	Moderate
7 to 9	High
10 +	Very high

Table 3.3 Ultraviolet index.

Ultraviolet radiation is used in the treatment of certain skin diseases such as psoriasis. This involves the use of a drug called psoralin being taken and then the patient being exposed to a carefully controlled amount of radiation.

Lasers in medicine

A laser is a very concentrated form of light. The light is also of one particular wavelength. Soon after the first laser was made in 1960 by Theodore Maiman, it was described as a 'solution looking for a problem'! They are now used in a wide variety of areas. There is a laser in a CD or DVD writer and a less powerful one in a CD or DVD player.

In medicine the laser has proved itself invaluable for some types of surgery, yet it has not replaced the scalpel to the extent it was predicted. In medicine, the laser is used to produce extreme heating in a very small piece of tissue. In one application, the laser beam is used to seal blood vessels by coagulating them. In another, the narrow beam is focused on a tumour, causing it to vaporise. The properties of different lasers are shown in table 3.4.

Laser	Power	Wavelength (nm)
Carbon dioxide	20W	10 600 (infrared)
Helium neon	5 mW	630 (red light)
Argon	I W	500 (blue-green)
Neodymium-YAG	50 W	1064 (infrared)
nm = nanometre = 1/1 000 000 000 m.		

Table 3.4 Properties of different lasers.

Doctors use the neodymium-YAG laser to vaporise tumours that obstruct the flow of air to the lungs. After laser treatment for a tumour that blocks the oesophagus, the patient can swallow again.

Laser as a scalpel

As the carbon dioxide laser beam is almost totally absorbed in the first tenth of a millimetre of tissue, it is particularly suited for use as a 'laser scalpel'. The shallow penetration makes it ideal for treating areas where it is important not to damage underlying structures. Certain malignant tumours can be vaporised using a carbon dioxide laser.

(a)

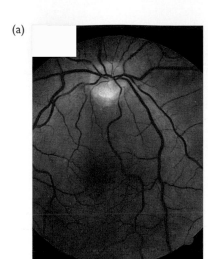

(b)

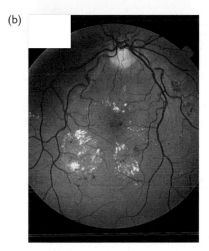

Figure 3.32 (a) The retina of a healthy person, (b) The retina of a diabetic before treatment for diabetic retinopathy.

Eye problems

Eye surgery is the best known application of argon lasers. The retina of a diabetic person sometimes does not get enough oxygen from the blood vessels. To compensate for this lack of oxygen, abnormal vessels grow forwards and bleed into the eye. Vision at the edge is altered and the patient can eventually go blind (figure 3.32).

The eye surgeon uses an argon laser to seal the less important areas of the retina. Although this reduces the patient's field of vision, the patient is much less likely to go blind.

This technique can be used for repairing retinal tears and holes which develop prior to the retina coming away from the back of the eye.

Hilly Janes
Observer

Sunday July 15, 2001

One morning before Christmas last year, 27-year-old Mark Weston woke up to find that he could see nothing through his left eye. A few months later his right eye started going. 'It was like looking through a bottle of milk, a really thick fog,' he recalls. 'I was more or less blind.'

After several operations the sight in his 'good' eye has been, for now at least, partially restored, although his left eye is irreparably damaged. But life is hard for Mark, who lives alone in north London and is unemployed. And he lives with the knowledge that not only is the preservation of what is left of his vision largely up to him, but also that the damage was self-inflicted.

Mark is suffering from diabetic retinopathy, a complication of diabetes and the most common cause of blindness among young people. He has been diabetic since he was nine, although the problems with his eyes only started about two years ago when he began to see spots before his eyes. If he had kept his diabetes under control and sought early treatment for his vision, he would not be in the state he is now.

'Unfortunately, many teenage diabetics don't consider control a priority and have very high blood sugars for many years. Then, when they are in their twenties, the complications hit, when it is too late

to avoid them,' says Bill Aylward, Mark's consultant ophthalmic surgeon at Moorfields Eye Hospital in London. Of the UK's 1.4 m diabetics, only 10 per cent will develop symptoms as severe as Mark's.

Mark needs to keep controlling his diabetes for the rest of his life and attend regular eye checkups. If he does, the prognosis for maintaining what sight he has left is reasonable. For now he is coping. 'It's quite difficult going out, even just for a walk, so I normally go with a friend,' he explains. 'I can see figures on the TV but not detail and I can't read at all.' He can also make simple meals for himself such as soup and toast, but cutting up food is tricky. Housework is difficult, too. Settings on the washing machine and cooker are hard to read and he has only just started vacuuming his flat again because he can see what he's supposed to be doing. Yet he tries to remain cheerful. 'I get out as much as I can,' he says. 'I listen to music and I've started doing weights. Friends come round and sit with me for a couple of hours and I try and laugh about it so I don't get too depressed, but it does get pretty boring.'

Can he stay motivated? 'I'm frightened of losing my eyesight completely. It's hard enough as it is at the moment,' he responds. 'I don't know what it would be like if it did happen. So, I try to live for the day. I can't mend all the years of damage I've done, but I can try to look after myself now. I've go to do it.'

Lasers and short sight

The most recent work with lasers involves sculpting the eye to remove the problem of short sight. A laser is used to change the curvature of the cornea. A mask is placed in front of the eye and the laser vaporises part of the cornea (figure 3.33). However, the change is permanent and about 10 per cent of patients seem to report problems. A different type of process is the LASIK technique in which an opening is made in the cornea to form a hinge for part of the cornea. A laser is used to vaporise part of the cornea (figure 3.34).

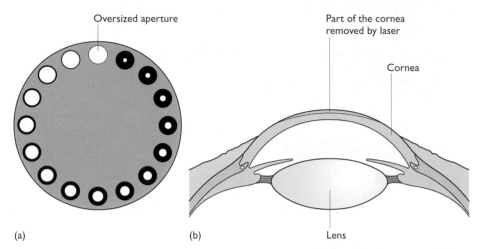

Oversized aperture

Part of the cornea removed by laser

Cornea

(a)

(b)

Lens

Figure 3.33 (a) Aperture wheel used in the correction of short sight, (b) effect on cornea of laser treatment.

Port wine birth marks are caused by blood vessels which have not sealed properly. The light from the argon laser is absorbed by the blood vessels causing them to seal. A similar treatment can be used on some tattoos. In this case the argon laser breaks up the dye (figure 3.35).

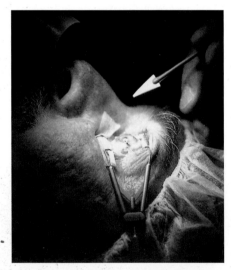

Figure 3.34 LASIK treatment being used to correct vision.

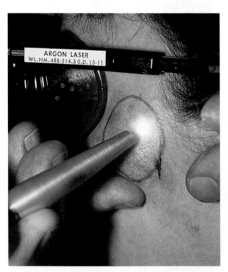

Figure 3.35 Treating 'port wine' birthmark with a laser.

A further use of lasers is when a patient has had a lens inserted in the eye after a cataract operation. Strands of tissue grow behind the eye which do not allow the light to pass through the lens. A few pulses from the neodymium-YAG laser will split the tissue and restore the patient's sight immediately.

Holograms

Holograms were first suggested by Dennis Gabor in 1948. It was not possible to make them until the invention of the laser. They are made by splitting a laser beam with a specially coated glass. The beam then travels two different paths, one of which reflects off an object. When the two paths meet on a photographic film they appear to consist of a set of black and white lines. But if the film is put back in a laser beam then the original object is seen and in three dimensions (3D). A popular use of a special type of hologram called a reflection hologram is used in credit and bank cards for security purposes.

Fascinating Physics

Medical X-rays

X-rays are used either to see inside the body or to treat some diseases. They are made by putting a high voltage across a tube which has been evacuated (no air is present). One of the wires is heated to emit electrons which are accelerated until they hit another piece of metal which gives off X-rays. The greater the voltage across the two plates the greater the energy of the X-rays. If the voltage is between 80 and 120 kV then the X-rays can be used to produce pictures. Voltages of greater than 200 kV can produce X-rays for treating cancers.

The use of X-rays in medicine depends on the fact that they pass through body tissues like skin, fat and muscle fairly easily, but are more readily absorbed by bones. When X-rays hit the photographic plate on the other side of the patient, they affect the photographic emulsion and blacken it, and so the image of the arm would be fairly dark, with lighter areas for the bone. The bones are white since they absorb the X-rays. The degree of blackening on the plate will depend on the number of rays reaching it. See figure 3.36.

The photographic plate can be placed between two other plates called intensifying screens which reduce the X-ray exposure required to produce a picture. Any break in a bone lets X-rays through and may show up as a dark crack (figure 3.37).

The boundary between organs, both made of similar tissue, will not be clear. A contrast agent is used, such as iodine or barium, which absorbs X-rays. Barium is drunk by the patient to outline the stomach and iodine is used to outline arteries and veins. The organ which has the contrast agent will show up lighter on the photographic plate (figure 3.38). Some X-ray machines do not use film to record results but use special detector tubes called image intensifiers to receive the rays and convert them into electrical signals. These can be converted into digital signals and displayed on a monitor screen (figure 3.39).

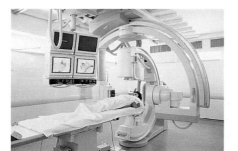
Figure 3.36 X-ray machine with digital display.

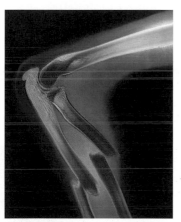
Figure 3.37 X-ray showing a bone fracture.

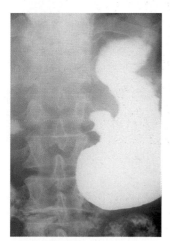
Figure 3.38 Using a contrast agent.

Figure 3.39 A digital angiogram.

Computed tomography

The problem

In a normal X-ray the details of organs at different depths are laid on top of each other and any information on depth is lost. You can try to take X-rays at different angles but this means that the radiologist has to work out which parts of each image correspond to each organ. This is a very difficult task to do, particularly with some organs. Normal X-rays cannot show small differences in the density of tissue. This is quite important in detecting disease.

The solution

A team of scientists at EMI made a machine called the CAT (computerised axial tomography), also called computed tomography. The machine uses an X-ray tube and a detector which are mounted on a gantry which can scan around the patient. The system takes a series of measurements in one position by scanning around the body. Then the patient is moved into a new position and a repeat series of measurements are taken. The process is then repeated for a series of slice thicknesses which are about 10 mm. A kidney CT scan would have 10 slices while others which have greater depth will require more (figure 3.40). The measurements are then used by a computer which can carry out 20 million operations. These will produce a slice image which is displayed on the screen (figure 3.41). Early CT scanners took up to 15 minutes to make measurements from a slice but the latest scanners can take the measurements in one to two minutes. This uses a fan shaped beam and a complete ring of 720 detectors (figure 3.42). If the patient moves while the measurements are taken then the image will be fuzzy.

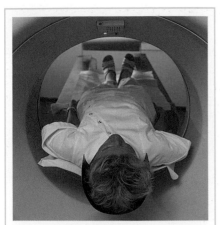

Figure 3.40 CAT scanner.

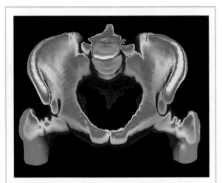

Figure 3.41 Slice image of a pelvis from a CT scan.

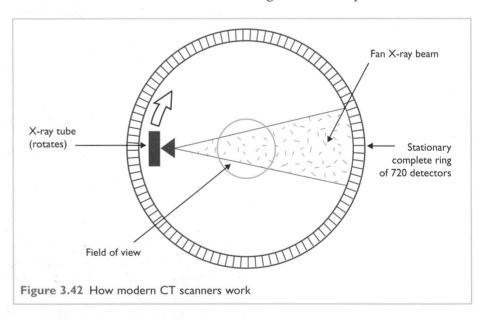

Figure 3.42 How modern CT scanners work

One main use of these CT scanners is by brain surgeons. The very small changes in tissue caused by cancer can be used to detect brain tumours. Another use is for detecting internal bleeding which can occur as a result of a car accident.

Figure 3. 43 shows the bleeding that occurred as a result of a sports injury. During a football match a player sustained an injury to the head that resulted in dizziness and a feeling of sickness. A CT scan showed that internal bleeding had occurred and knowing the exact location of the bleeding will allow the surgeon to operate and drain away the blood.

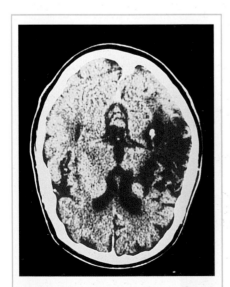

Figure 3.43 A CT scan showing an internal bleed in the brain.

End of Section Questions

1 Give a use of lasers in medicine.

2 During a building collapse some people are suspected of being trapped. They can be detected by the heat given out by their bodies. What kind of radiation is being detected?

3 We are constantly exposed to ultraviolet radiation from the sun.

(a) Why is it necessary to receive some of this type of radiation?

(b) Excess exposure to this radiation can cause problems. What might happen if this occurs?

4 When you have a suspected broken limb, you may be given an X-ray.

(a) What is used to detect these rays?
(b) Explain why a break is shown as a white or a black line.

5 To detect a problem with possible internal bleeding, a CT scan is taken. What is the advantage of this technique?

SECTION 3.5 The atom and radiation

At the end of this section you should be able to;

1 State that radiation can kill living cells or change the nature of living cells.
2 Describe one medical use of radiation based on the fact that radiation can destroy cells e.g. (instrument sterilisation, treatment of cancer).
3 Describe one medical use of radiation based on the fact that radiation is easy to detect.
4 State the range and absorption of alpha, beta and gamma radiation.
5 State that radiation energy may be absorbed in the medium through which it passes.
6 Describe a simple model of the atom which includes protons, neutrons and electrons.
7 State that alpha rays produce much greater ionisation density than beta or gamma rays.
8 State one example of the effect of radiation on non-living things (e.g ionisation, fogging of photographic film, scintillations).
9 State that the activity of a radioactive source is measured in bequerels.
10 State that the activity of a radioactive source decreases with time.
11 Describe the safety precautions necessary when dealing with radioactive substances.
12 State that the dose equivalent is measured in sieverts.

13 Explain the term ionisation.
14 Describe how one of the effects of radiation is used in a detector of radiation (e.g. GM tube; film badges; scintillation counters).
15 Describe a method of measuring the half-life of a radioactive element.
16 State the meaning of the term 'half-life'.
17 Carry out calculations to find the half-life of a radioactive element from appropriate data.
18 State that, for living materials, the biological effect of radiation depends on the absorbing tissue and the nature of the radiation and that dose equivalent, measured in sieverts, takes account of the type and energy of radiation.

Atoms

Particle	Charge
Proton	Postive
Neutron	Zero
Electron	Negative

Table 3.5 Particles making up an atom.

Atoms are the smallest possible particles of the elements which make up everything around us. All atoms of the one element are identical to one another, but they are different from all other elements. This is because they are made up from different combinations of electrons, protons and neutrons – the three main particles which make up atoms (table 3.5).

All atoms have a tiny central nucleus which has a positive charge. We can imagine the negatively charged electrons to be circling around this, rather like planets around the sun. The nucleus contains the positive protons and the neutrons, which are uncharged. Look back at figure 2.7 on page 35.

Radiation

There are three types of radiation:
◆ Alpha (α) particles;
◆ Beta (β) particles;
◆ Gamma (γ) rays.

These can be identified by what happens as they reach different materials. Alpha radiation can be completely stopped by a few sheets of paper. Beta radiation can be absorbed by a sheet of aluminium. Gamma radiation is part of the electromagnetic spectrum. Concrete, lead and other dense materials will absorb gamma radiation. The thicker the material the more radiation that will be absorbed.

When the alpha or beta or gamma radiation pass through a material they lose energy by colliding with the atoms of the material. Eventually the radiations lose so much energy that they cannot get through (penetrate) the material and so are absorbed (figure 3.44).

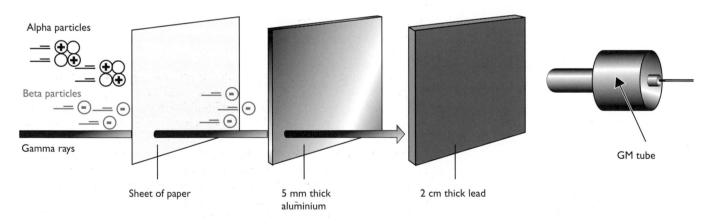

Figure 3.44 Absorption of different radiations.

Ionisation

If an electron is added or removed from an atom, what is left is called an **ion**. The process is called **ionisation**. The removal of an electron creates a positive ion and if an electron is added then a negative ion is formed.

The process of ionisation by an alpha particle is shown in figure 3.45. In (a) the alpha particle is approaching the neutral atom and in (b) it has passed by, having created an ion pair. This means that the alpha particle has caused the atom to break up into positive and negative parts called ions.

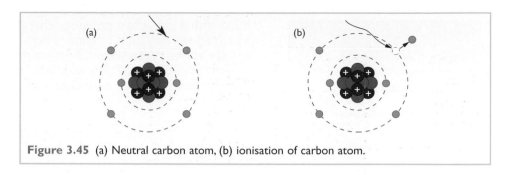

Figure 3.45 (a) Neutral carbon atom, (b) ionisation of carbon atom.

Effect on the human body

◆ Alpha radiation will produce ionisation in a short distance of body tissue. This type of radiation outside the body is absorbed by the skin and little damage will occur. Swallowing the radiation will produce large amounts of ionisation and would be dangerous.

◆ Beta radiation will be absorbed in about one centimetre of tissue and any beta radiation outside the body will cause damage to that tissue but a small amount can penetrate the body. If the radioactive source were to get into the body then internal organs could be damaged.

◆ Gamma rays and X-rays pass through the body and can damage tissue whether the source is inside or outside of the body.

Sterilisation

As radiation can be used to kill cells, it can also be used to kill bacteria or germs. In the past, medical instruments such as syringes had to be sterilised by heat or chemicals. Now cheap, plastic, throwaway syringes can be used – they are prepacked and then irradiated by an intense gamma ray source. This kills any bacteria but does not make the syringe radioactive.

Detecting radiation

Photographic fogging

Photographic film has a thin layer of silver-based chemical on the surface of the plastic or paper. Normally, this silver salt is affected by light falling on it – wherever it lands, it changes the chemical and blackens or fogs the film surface.

Alpha, beta or gamma rays have a similar effect on this photographic emulsion, and so photographic film can be used to detect them. In fact, radioactive substances were first discovered – by accident – when Henri Becquerel left some uranium rocks near photographic paper. He discovered that the paper had been blackened, and when he went on to investigate why, he started the study of radioactivity.

Workers who use radioactive materials – for example health workers in hospitals – wear film badges throughout their working day so that a check can be made on the amount of radiation they have been exposed to. When the film is developed, the amount of fogging gives a measure of the radiation exposure (figure 3.46). Different windows are used to measure the amounts of the different types of radiation.

◆ A plastic window will absorb different energies of beta rays.
◆ Metal windows absorb different energies of gamma and X-rays .
◆ Aluminium will absorb low energy X-rays.
◆ The other metals will absorb the high energy X-rays.

Geiger Muller (GM) tube

The most common method of detecting radiations in the school laboratory is a GM tube. This was developed by German scientists working in the Cavendish Laboratory in Cambridge in the early 1900s. It is a hollow tube

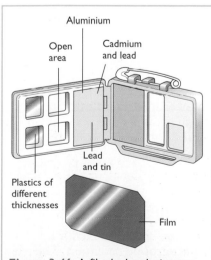

Figure 3.46 A film badge dosimeter worn on the worker's waistband below protective clothing, such as a lab coat.

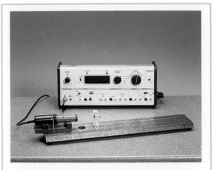
Figure 3.47 A Geiger Muller tube connected to a digital counter.

filled with a gas at low pressure. There is a thin window of mica at one end which allows radiation to enter. There are two electrodes at one end with a voltage across them. When radiation passes through the window, it causes ionisation in the gas. The ions produce electrical pulses which can be counted and displayed on a digital counter (figure 3.47).

Scintillations

Some substances such as zinc sulphide are fluorescent. This means that they absorb radiation and give out energy again as a tiny burst of light. These flashes of light are called 'scintillations', and they may be observed by the naked eye or counted by a light detector and an electronic circuit. These scintillation counters are used in many modern instruments including the gamma camera.

Radioactive decay

Radiation from a source is caused by the radioactive atoms breaking up. The activity of a radioactive source is a measure of how much radiation it is giving out. This depends on the number of radioactive atoms which break up every second and give out radiations. The unit of **activity** is the **becquerel** (**Bq**). A source has an activity of 1 Bq if one of its atoms disintegrates each second and gives out a particle of radiation. The becquerel is a very small unit. Radioactive sources used in medicine have activities measured in megabecquerels (MBq).

$$\text{One million Bq} = 1\,\text{MBq} = 10^6\,\text{Bq}$$

Half-life

When a radioactive substance disintegrates the activity (number of disintegrations or count rate) depends only on the number of radioactive nuclei present, that is double the number, double the activity.

The half-life of a radioactive substance is the time taken for half the radioactive nuclei to disintegrate that is the time taken for the activity to fall by one half.

Half-life is measured in units of time – seconds, minutes, days or years. Typical half lives are:

- Uranium-238: 4.5×10^9 years
- Radium-226:   1600 years
- Cobalt-60:   5.3 years
- Sodium-24:   15 hours
- Copper-66:   5.2 minutes

After one half-life ($t_{1/2}$), the activity and the measured count rate drop to half the initial value. After a second half-life, the count rate halves again – it is now one quarter of its original value. Three half-lives will see the activity and count rate reduce to 1/8th, and so on.

A graph of count rate against time is shown in figure 3.48. Taking the initial activity as 1, a table can be produced (table 3.6).

Total count rate and background count rate

In any radioactive experiment a count rate will be obtained even if no radioactive source is present. This is due to background radiation. The background count rate could be found by measuring the number of background counts in a known time and working out an average count

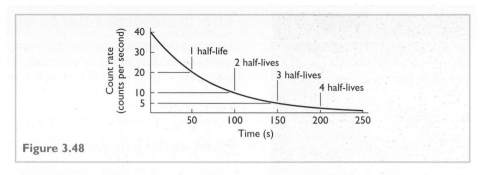

Figure 3.48

Number of half-lives	0	1	2	3	4
Activity	1	1/2	1/4	1/8	1/16

Table 3.6

rate from this. Alternatively, when carrying out your experiment you could measure the total count rate with your measuring device where:

Total count rate = count rate from radioactive source + count rate due to background

Example
A radioactive source gives an initial count rate of 1600 Bq.
After 120 minutes the count rate is found to be 100 Bq.
Calculate the half-life of the source.

Solution

> After 1 half-life the count rate is 800 Bq
> After 2 half-lives the count rate is 400 Bq
> After 3 half-lives the count rate is 200 Bq
> After 4 half-lives the count rate is 100 Bq
> 4 half-lives = 120 minutes
> 1 half-life = 30 minutes.

Example
A radioactive source has a half-life of 4 days. At the start of an experiment the total activity recorded is 990 counts per minute. Find the total recorded activity after 20 days, if the background count rate is 30 counts per minute.
Note:
1   Total count rate = count rate of source + background count rate.
2   The background rate is constant and does not decrease with time.

Solution

> Count rate of source = 990 − 30
> = 960 counts per minute
> Number of half-lives = $\frac{20}{4}$
> = 5
> After 1 half-life the count rate is 480 counts per minute
> After 2 half-lives the count rate is 240 counts per minute
> After 3 half-lives the count rate is 120 counts per minute
> After 4 half-lives the count rate is 60 counts per minute
> After 5 half-lives the count rate is 30 counts per minute
> Total recorded count rate is 30 + 30 = 60 counts per minute

Figure 3.49 Chernobyl nuclear power station near Kiev, Ukraine.

The biological effects of radiation

All living things, plant or animal, are made of cells. Ionising radiation may damage the cells they pass through. The damage caused may be severe and cause immediate effects, or it may be more subtle and have effects which are not seen for a long time. The effects depend on both:

◆ the type of radiation, and
◆ the part of the body the radiation is going through.

Short-term effects

On 26 April 1986 there was an accident at the Chernobyl nuclear power station near Kiev in Russia. A fire there caused a number of firemen to be exposed to very large amounts of radiation, and around thirty died as a result (figure.3.49). The explosion was caused by an experiment on the reactor going wrong. Over 30 fires were started and many firefighters lost their lives due to the radioactive materials emitted from the core of the reactor. The damage to the immediate area was extensive but the radiation effects over a wide area were considerable: 135 000 people were removed from an area which had a radius of 30 km. The smoke and radioactive debris reached a height of 1200 m and travelled across Russia and Poland and then to Scandinavia. On 2 May 1986 the cloud of material reached Great Britain and with heavy rain there was material deposited on parts of north Wales, Cumbria and Scotland. This caused certain farm animals, such as lambs, to be banned from sale since they absorbed products of radioactive fission from the grass.

Long-term effects

There are other effects of radiation which take much longer to show. In some ways these are more important to us because they can be caused by much lower levels of radiation. The most important long-term effect is to cause cancers in various parts of the body. There is a lot of evidence of these effects. Uranium miners tended to get lung cancer due to breathing in gases which emitted alpha particles. People who painted the dials of clocks with luminous paint developed bone cancer from using their lips to make points on the brushes. Marie Curie and many other early workers with radioactivity died of forms of cancer.

Genetic damage can be caused to cells, including the cells which are involved in reproduction. Plant and animal studies have shown small increases in the numbers of mutations in future generations. These mutations – changes to the structure of the plant or animal – are usually harmful. But they are the ones which occur naturally anyway, although the rate of mutation is greater after the irradiation.

The total effect of the radiation is a combination of:

◆ The nature of the radiation.
◆ The type of body tissue which absorbs it.
◆ The total amount of energy absorbed.

Equivalent dose

The biological risk caused by the radiation is represented by a quantity called the **equivalent dose**. It is measured in a unit called the sievert (Sv), although we often talk of doses in millisieverts (mSv) or microsieverts (μSv). (A millisievert is a thousandth and a microsievert a millionth of a sievert.) A sievert is a very large dose of radiation, and could only happen as a result of a very serious accident or after a nuclear explosion.

How dangerous is an equivalent dose of 1 Sv? It is impossible to say for any one person. However, suppose that 100 people all receive a dose equivalent of 1 Sv spread over the whole body. It is estimated that, of the 100 people, on average 4 of them would eventually die as a result of the radiation. But precisely who would die, or when they would die, or what illness they would die of, cannot be predicted.

Background radiation

We have seen that there is radiation all around us – our detectors in the lab pick up radiations even when none of the lab sources is anywhere near. This is known as background radiation, and it is almost all from natural sources. Tables 3.7 and 3.8 below show the typical equivalent dose we get every year from background radiation and other sources. They are average figures, and vary a lot depending on our job and where we live. The different percentages are shown in figure 3.50.

Source	Annual dose (μSv)
Radon and thoron gas from rocks and soil	800
Gamma rays from ground	400
Carbon and potassium in body	370
Cosmic rays at ground level	300
	Total = 1870

Table 3.7 Natural sources of radiation.

Source	Annual dose (μSv)
Medical uses – X-rays, etc.	250
Chernobyl (first year)	50
Fall-out from weapons testing	10
Job (average)	5
Nuclear industry (e.g. waste)	2
Others (TV, aeroplane trips, etc.)	11
	Total = 328

Table 3.8 Man-made sources of radiation.

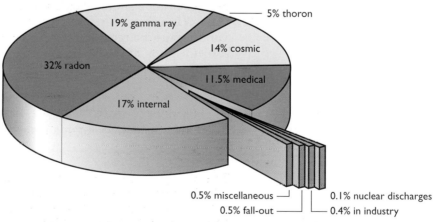

Figure 3.50 Average percentage sources of annual dose of radiation to the population of the UK.

You can see from the tables that natural radiation is far the biggest influence on us. The total annual equivalent dose in the UK averages about 2000 μSv, or about 2 mSv. But there is a big variation from person to person. If you take several flights across the Atlantic each year you will have a greater risk. If you live in Aberdeen or especially in Cornwall, where the granite rocks are giving off radioactive radon gas, you are subjected to a much higher background rate. Part of the time allocation to pilots and other staff for flying takes account of the exposure to radiation.

Other risks

It is interesting to compare radiation risks with those of other types. It is estimated that medical uses of radiation – while they can have great value and save many lives – will probably cause the death of one person out of every 240 000. Other risks are given in table 3.9 for comparison.

Death risk – cause	Death risk for a 40 year old
All causes	1 per 500
Smoker – 10 cigarettes per day	1 per 2000
Road accidents	1 per 5000
Home accidents	1 per 10 000
Work accidents	1 per 20 000
All radiations	1 per 27 500
Medical radiations	1 per 240 000

Table 3.9 Radiation risks compared with other risks.

There are people who, as a result of their work, are exposed to ionising radiations. These people include medical workers using X-rays in hospitals, dentists and vets, research workers using radiation sources in their experiments and people who work in the nuclear power industry. Shielding a source of radiation with an appropriate thickness of absorber can reduce the risk. For example, a radiographer wears a lead-lined apron.

Safety with radioactivity

1 Always use forceps or a lifting tool to remove a source. Never use bare hands.
2 Arrange a source so that its radiation window points away from the body.
3 Never bring a source close to your eyes for examination.
4 When in use, a source must always be attended by an authorised person and it must be returned to a locked and labelled store in its special shielded box immediately after use.
5 After any experiment with radioactive materials, wash your hands thoroughly before you eat.
6 In the UK, students under 16 years old may not normally handle radioactive sources.

The symbol that shows radiation sources are being stored is shown in figure 3.51.

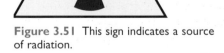

Figure 3.51 This sign indicates a source of radiation.

Using radiation in medicine

Treating cancer – radiation therapy

Radiotherapy is the treatment of cancers by radiation. Cancers are growths of cells which are out of control. Cancerous tumours can be treated by drugs, surgery or radiation. The choice of treatment depends on the size and position of the tumour. The object of the radiation treatment is to cause damage to the cancer cells which then stop reproducing. The tumour then shrinks.

Unfortunately, healthy cells can also be damaged by radiation, and so the amount of radiation has to be very accurately calculated so that sufficient damage is done to cancer cells without overdoing the damage to other cells.

The radiation must be aimed very accurately at the tumour. This can be done using a simulator. A series of X-ray photographs are taken at different angles and a computer can build up a picture of the tumour and measure the amount of radiation to be given. Some localised tumours (e.g. a bone tumour) can be treated by irradiation with high energy X-rays or gamma rays (figure 3.52)

The gamma rays are emitted from a Cobalt-60 source – a radioactive form of cobalt. The cobalt source is kept within a thick, heavy metal container. This has a slit in it to allow a narrow beam of gamma rays to emerge.

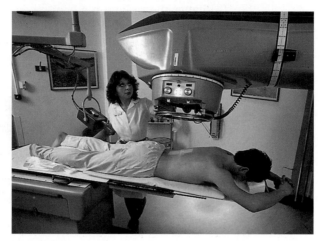

Figure 3.52 Gamma radiotherapy being used to treat bone cancer.

The X-rays are generated by a linear accelerator. This machine fires high energy electrons at a metal target and when the electrons strike the target X-rays are produced. The X-rays are shaped into a narrow beam by movable metal shutters.

With either technique, the apparatus is arranged so that it can rotate around the couch on which the patient lies. This means the patient can receive radiation from different directions. This allows the diseased tissue to receive radiation all of the time but the healthy tissue receives only a small amount of radiation. Treatments are given as a series of small doses because tumour cells are killed more easily when they are dividing, and not all cells divide at the same time. This reduces the side effects, such as sickness.

The gamma camera

It is important for scientists to be able to study internal organs without surgery. To see how the kidneys are working a **radiopharmaceutical** is used which can act as a tracer. The advantage of the radiopharmaceutical is that the radiation can be detected outside the body.

The radiopharmaceutical has two parts (figure 3.53):

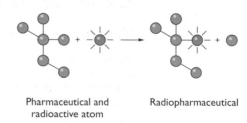

Pharmaceutical and radioactive atom   Radiopharmaceutical

Figure 3.53 Formation of radiopharmaceutical.

♦ A drug which is chosen for the particular organ that is being studied. (Different organs have different drugs.)
♦ A radioactive substance which is a gamma emitter.

Gamma is chosen since alpha or beta would be absorbed by tissue and would not be detected outside the body. This is normally technitium 99m which is used because it has a half-life of 6 hours. The half-life is important because:

♦ For a shorter time than six hours it would be too difficult to make measurements.
♦ A longer time would increase the amount of radiation to the body.

The combined substance is then injected into the patient.

To detect the radiation a gamma camera is used. This device has special crystals which give off flashes of light when the radiation reaches them. (The effect is called scintillation.) Tubes called photomultipliers can change the light energy into electrical energy. The signals can then be displayed on a screen (figure 3.54).

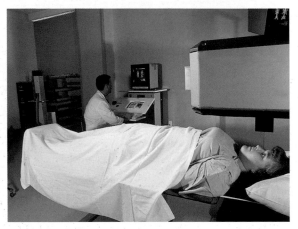

Figure 3.54 Gamma camera.

There are two types of studies:

♦ A static study, where there is a time delay between injecting the radioactive material and the build up of radiation in the organ. This occurs in bone, lung or brain scans (figure 3.55).
♦ A dynamic study, where the amount of radioactivity in an organ is measured as time increases. This occurs in the examination of the operation of the kidneys.

Kidney examination or renogram

This technique examines the working of the kidneys. The radioactive material reaches the kidneys due to the drug given to the patient. The radioactive material is taken out of the bloodstream by the kidneys. Within a few minutes of the drug being injected the radiation is concentrated in the kidneys. After 10 to 15 minutes, almost all the radiation should be in the bladder. The gamma camera takes readings every few seconds for 20 minutes. The computer adds up the radioactivity in each kidney. This can be shown as a graph of activity against time (figure 3.56). The left kidney is not working correctly since the radioactivity is taken up by both kidneys but does not decrease rapidly as happens with the other kidney.

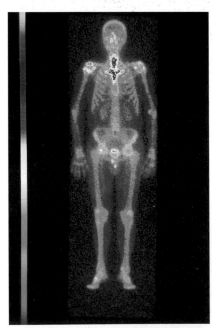

Figure 3.55 A gamma scan of the human skeleton.

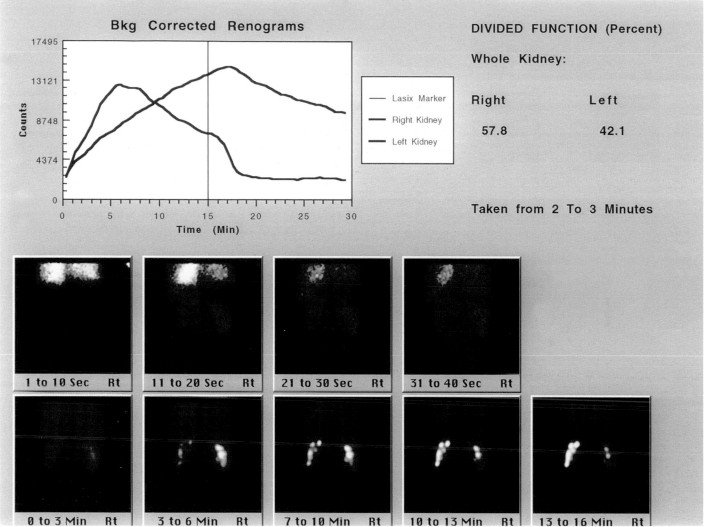

PATIENT NAME : EXAMPLE2   INSTITUTE : RADIOISOTOPE SERVICE MEDICAL PHYSICS LEI
PATIENT ID   : F2820RD   PROTOCOL : RENOGRAM
BIRTH DATE   : 1975.08.28   ACQ. DATE : 11-NOV-1996

Figure 3.56 Renogram showing different levels of activity in each kidney.

Magnetic resonance imaging

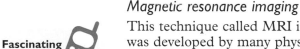

Fascinating Physics

This technique called MRI is used to see into the body without radiation. It was developed by many physicists including some in Aberdeen University. The technique is based on the idea that the nuclei of certain atoms behave as small atoms due to the fact that they spin.

When the human body is placed in a very powerful magnetic field, about half of the nuclei line up in the direction of the field and about half line up in the opposite direction. There is then a pulse of high frequency radio wave, which causes the tiny magnets to change direction and absorb energy at particular frequencies, called resonant frequencies (RF), which depends on the strength of the magnetic field (figure 3.57). When the RF pulse ends, the nuclei return to their original state. This process is called 'relaxation' and the times for this process can be measured. Hydrogen in water tissue has long relaxation times but in fat it has short relaxation times. This allows different tissues to be seen (figure 3.58). A typical scan is shown in figure 3.59.

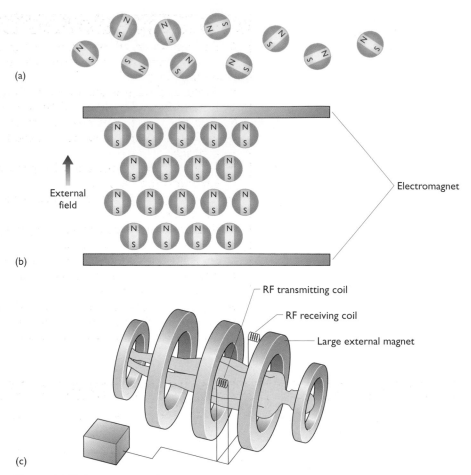

(a)

(b)

External field

Electromagnet

RF transmitting coil

RF receiving coil

Large external magnet

(c)

Figure 3.57 Apparatus used in magnetic resonance imaging.

Advantages of MRI are:
- ◆ No ionising radiation is used.
- ◆ It is better for displaying soft tissue than a CT scan.
- ◆ There are no after effects.

Disadvantages of MRI are:
- ◆ Any metallic objects such as surgical pins for fractures or clips will heat up.
- ◆ Pacemakers are affected by the magnetic fields.

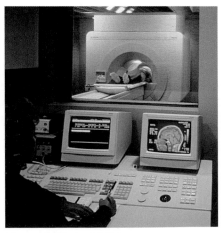

Figure 3.58 An MRI Scanner.

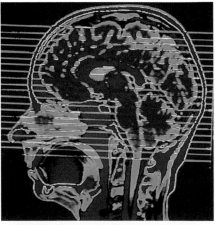

Figure 3.59 An MRI Scan.

Section 3.5 Summary

◆ Radiation can kill or change living cells.
◆ This change in cell structure can be used to sterilise instruments or treat cancer.
◆ Alpha radiation is absorbed by 5 cm of air or a piece of paper.
◆ Beta radiation is absorbed by a thin piece of aluminium.
◆ Gamma radiation is reduced by concrete or lead.
◆ Radiation energy may be absorbed in the substance which it passes through.
◆ The atom has a nucleus with protons and neutrons. Electrons surround the nucleus.
◆ Alpha rays produce much greater ionisation density than beta or gamma rays.
◆ Radiation can cause ionisation, fogging of photographic film and scintillations in materials.
◆ Ionisation is the gain or loss of an electron to leave a charged particle.
◆ The activity of a radioactive source is measured in becquerels and decreases with time.
◆ Equivalent dose is measured in sieverts.
◆ Half-life is the time for the activity to halve from its initial activity.

End of Section Questions

1 A radioactive source gives out two different types of radiation. Radiation A passes through paper but is absorbed by aluminium. Radiation B passes easily through the air and is reduced by a piece of lead. Identify the two radiations.

2 Alpha sources of radiation produce large amounts of ionisation.

   (a) What is meant by ionisation?
   (b) This ionisation rarely causes problems in air but when the source is placed in the body there can be difficulties due to ionisation density. Suggest why this is a problem.

3 State the units of measurement of (a) activity and (b) equivalent dose. State one factor on which the equivalent dose depends.

4 Gamma radiation is dangerous but it is used to detect problems using a gamma camera. Why is a gamma source used rather than other sources?

5 State two precautions in handling radioactive sources.

6 Radioactive sources can cause different effects on substances. What is meant by scintillations?

7 (a) What is meant by half-life of a radioactive source?
   (b) A source has an activity of 256 MBq. When the activity is measured in 2 days it is found to be 4 MBq. Calculate the half-life of this source.

8 (a) What is meant by background radiation?
   (b) A source has a total measured activity of 1250 Bq on a day when background radiation is measured as 50 Bq. If the half-life is 4 hours, what will the measured activity be in 12 hours?

1   During a mountain climb, Angus gets lost when the mist comes down suddenly. He checks his temperature with a pocket clinical thermometer.

(a) Describe the key features of a liquid-in-glass clinical thermometer.
(b) His temperature is measured at 34°C. Explain why this temperature is important.
(c) After some time he is rescued and is checked for injuries. He has an X-ray taken to check for broken limbs.
   (i) What is used to detect the X-rays?
   (ii) As an additional check a CT scan is taken. What is the advantage of this technique?

2   During a visit to the optician, Carole finds that she is short-sighted.

(a) What is meant by short sight?
(b) What type of lens is needed to correct this eye condition?
(c) The focal length of the correcting lens is 15 cm. Calculate the power of this lens.

3   Ultrasound is often used in examinations.

(a) What is meant by ultrasound?
(b) A frequency of 5 MHz is used to examine tissue. The speed of sound in tissue is 1500 m/s. Calculate the wavelength of the sound.
(c) The smaller the wavelength the greater the detail that can be seen, but the more energy that is absorbed. Explain why a frequency of 3 MHz is better to use for an organ which is much thicker than thin tissue.

4   For part of a fitness programme, some athletes are exposed to some ultraviolet light by special lamps.

(a) Why is some ultraviolet light necessary for the body?
(b) What is the danger of excessive exposure to ultraviolet light?
(c) As a result of an injury one athlete is given heat treatment using a radiation.
   (i) What is the name for this radiation?
   (ii) How does its wavelength compare with that of ultraviolet?

5   Radiation is often used to detect possible illnesses in the human body.

(a) Gamma radiation combined with a drug is used as a tracer to detect problems with blood flow. Explain why a gamma emitter is used rather than an alpha or beta emitter.
(b) The gamma emitter has a half-life of six hours. A student suggest that a half-life of 15 minutes would be suitable since that is the time for the measurements to take place.

Explain why such a half-life is not suitable for this investigation.

(c) The activity of the radiation is 50 kBq when inserted into the blood stream. If the half-life is 6 hours what will the activity be after 24 hours?

6   The endoscope is used to see inside the body by using fibre optics.

(a) What is an optical fibre?
(b) What is the advantage of using this device to see inside the body rather than an ordinary light source?
(c) Draw a diagram to illustrate the passage of light down an optical fibre.

CHAPTER FOUR
Electronics

Electrical systems

At the end of this section you should be able to:

1 State that an electronic system consists of three parts; input, process and output.
2 Distinguish between digital and analogue outputs.
3 Identify analogue and digital signals from waveforms viewed on an oscilloscope.

Figure 4.1 The picture shows a person using a loudhailer. Why do you think he needs to use a loudhailer?

Electronics is the science that deals with the control of electrons in an electrical circuit or electrical system. It usually involves the use of special electrical components such as 'transistors' or 'integrated circuits' (silicon chips). An **electrical system** is a collection of electrical components connected together to perform a particular function, for example a loudhailer (figure 4.1).

A simplified view of a loudhailer system would be:

◆ sound waves are changed into weak electrical signals;
◆ the weak electrical signals are then amplified (made bigger);
◆ the amplified electrical signals are then converted into sound.

Notice that it is convenient to break the loudhailer system into three parts:

1 the microphone which picks up the sound waves and converts them into weak electrical signals – known as the **input**;
2 the amplifier which boosts the weak electrical signals – known as the **process**;
3 the loudspeaker which converts the electrical signals to sound – known as the **output**.

In fact, all electronic systems can be broken into these three parts – input, process and output. The input section starts the system working, the process section alters the input so as to produce the required output, and the output section gives the desired result.

Since all electronic systems need to use electrical signals, devices are required to convert one form of energy (e.g. light, heat, sound, etc.) into electrical energy at the input stage and to do the opposite at the output stage.

Devices which convert input signals to electrical signals or electrical signals back to output signals are called **transducers**. Input transducers convert one form of energy into electrical energy, for example a microphone converts sound energy into electrical energy. Output transducers convert electrical energy into another form of energy, for example a loudspeaker converts electrical energy into sound energy.

An electronic system may be drawn as a 'block diagram' (figure 4.2). The arrows show how information is passed (electrically) from one block to another.

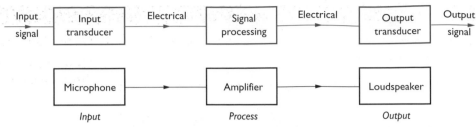

Figure 4.2 Block diagram for a loudhailer.

Analogue signals and digital signals

The signals used by electronic systems are of two types: **analogue** or **digital**. Figure 4.3 shows the traces displayed on an oscilloscope screen with (a) an analogue signal and (b) a digital signal.

Figure 4.3(a) shows a typical electrical signal from the microphone of a telephone when a person is speaking. The trace has a continuous range of values. This type of signal is called an analogue signal. Most input transducers produce analogue signals.

Figure 4.3(b) shows a typical electrical signal from a compact disc player. The trace has a series of electrical pulses each with the same amplitude. This type of signal is called a digital signal. In a digital signal the trace is either at a maximum value (called a **high** or **logic '1'**), or a minimum value (called a **low** or **logic '0'**).

An analogue signal has a continuous range of values, while a digital signal can have only one of two possible values.

Many electrical systems consist of both digital and analogue signals. An analogue signal produced by an input transducer may be converted into a digital signal in the process unit. For example, many telephone systems change the human voice (an analogue signal) into a digital signal – which can be transmitted over long distances – then back into an analogue signal (sound) that can be heard.

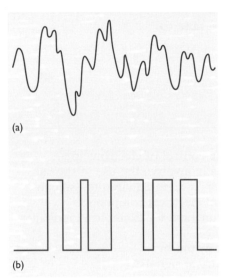

Figure 4.3 (a) Analogue signal, (b) digital signal.

Section 4.1 Summary

◆ All electronic systems can be broken down into three parts – input, process and output.
◆ An analogue signal has a continuous range of values.
◆ A digital signal has two possible values. The signal is either at a maximum value, called a high or logic '1', or at a minimum value, called a low or logic '0'.

1 State which of the signals in the figure below are digital and which are analogue.

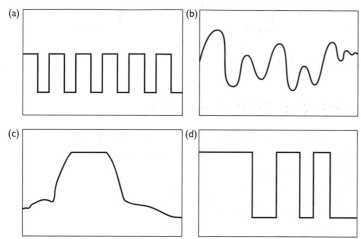

(a)

(b)

(c)

(d)

Figure 4.1Q1

2 List the following devices as analogue or digital:

Cassette recorder; CD player; computer; radio; television; thermometer (electronic); thermometer (mercury).

At the end of this section you should be able to:

1 Give examples of output devices and the energy conversions involved.
2 Give examples of digital output devices and of analogue output devices.
3 Draw and identify the symbol for an LED.
4 State that an LED will light only if connected one way round.
5 Explain the need for a series resistor with an LED.
6 State that different numbers can be produced by lighting appropriate segments of a seven-segment display.

7 Identify appropriate output devices for a given application.
8 Describe by means of a diagram a circuit which ones will allow an LED to light.
9 Calculate the value of the series resistor for an LED.
10 Calculate the decimal equivalent of a binary number in the range 0000–1001.

There are a number of output devices available for different applications.

The loudspeaker

A loudspeaker is an analogue output device which changes electrical energy into sound energy. A radio and a television are examples of electronic systems that contain a loudspeaker.

The electric motor

An electric motor is an analogue output device which changes electrical energy into kinetic (movement) energy. Its speed increases as the voltage is increased. The direction of rotation can be reversed by reversing the connections to the d.c. power supply. Vacuum cleaners and washing machines contain mains-operated electric motors.

Note: an electric motor could be a digital output device by connecting it to a battery and a switch. The motor would then be either 'on' when the switch was closed, or 'off' when the switch was opened.

The relay

A **relay** is a switch operated by an electromagnet. A coil of wire, when carrying an electric current, provides the magnetic field required to close the switch contacts in the relay shown in figure 4.4. When switch S is closed a current passes through the coil surrounding the switch. The switch contacts close completing the lower electrical circuit, thus allowing the lamp to light. When S is opened, the switch contacts open and the lamp goes out. The advantage of a relay is that a small current in one circuit is able to control another circuit containing a device such as a lamp, electric bell or motor which requires a larger current. The relay is a digital output device which changes electrical energy into movement energy – the opening or closing of a switch.

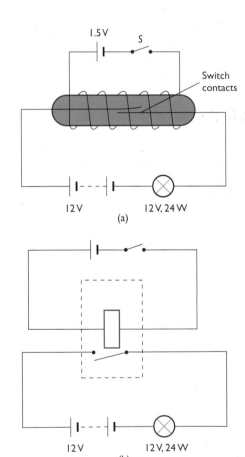

Figure 4.4 (a) A relay circuit – closing switch S allows the lamp to light, (b) circuit diagram for relay.

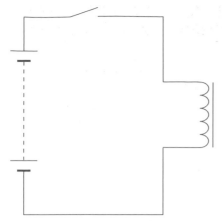

Figure 4.5 A voltage supply connected to a switch and a solenoid.

The solenoid

A **solenoid** consists of a coil of wire surrounding a metal core. When no current passes through the coil, a spring pushes the metal core away from the coil. However, when a large enough current passes through the coil, the magnetic field produced attracts the metal core into the coil.

Figure 4.5 shows a solenoid connected to a switch and a battery. When the switch is closed the metal core moves into the coil and is then held there. When the switch is opened, the metal core moves out of the coil and then stops. A solenoid is a digital output device which changes electrical energy into movement in a straight line. Solenoids are used in the central locking system of a car.

The filament lamp

A **filament lamp** consists of a thin tungsten wire (filament) in a glass container. When an electric current passes through the wire, electrical energy is changed into heat in the filament. A lamp is connected to a variable d.c. voltage supply. Increasing the voltage across the lamp, increases the current through it and so it gets brighter. No difference is observed when the connections from the d.c. power supply to the lamp are reversed. The filament in the lamp requires a relatively large current to light properly and gets very hot in operation. Lamps can be used as analogue or digital output devices. They are analogue devices if used with a dimmer circuit (i.e. the brightness changes) and digital devices if they are switched 'on' or 'off'.

The light-emitting diode (LED)

Light-emitting diodes are made by joining two special materials together to produce a junction. When an electric current passes through the junction it emits light. LEDs are available in red, green, yellow, blue and white colours. A series resistor must be used to limit the current or the junction will be destroyed. Figure 4.6 shows an LED connected to a variable d.c. voltage supply.

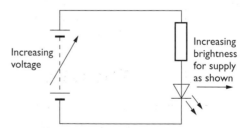

Figure 4.6 A variable voltage supply connected to a resistor and an LED.

Increasing the voltage across the LED increases its brightness. The LED does not light if the connections to the d.c. power supply are reversed. The LED only requires a small current to operate properly and does not get hot in operation. LEDs are usually used as digital output devices, i.e. either 'on' or 'off'. LEDs can be used in hi-fis and instrument panels.

Example
The maximum voltage allowed across an LED is 2.3 V and the current through it must not exceed 10 mA. The LED is connected to a 5 V d.c. supply. Calculate the value of the resistor, R, connected, in series, with the LED.

Solution
Since the LED and resistor are connected in series then $V_S = V_{LED} + V_R$ and the current through both components is the same (10 mA = 0.010 A).
Therefore:

$$V_R = V_S - V_{LED} = 5 - 2.3 = 2.7 \text{ V}$$
$$V_R = IR$$
$$2.7 = 0.010 \times R$$
$$R = \frac{2.7}{0.010} = 270 \ \Omega$$

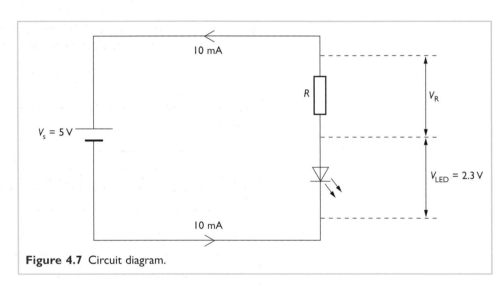

Figure 4.7 Circuit diagram.

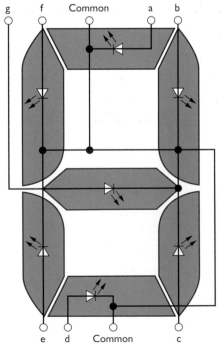

Figure 4.8 Part of seven-segment display showing each of the seven LEDs a to g.

The seven-segment display

A seven-segment display consists of 7 LEDs arranged in a rectangular package as shown in figure 4.8. Any number in the range 0 to 9 can be produced by lighting a number of the individual LEDs. For example, the number 1 is displayed by lighting LEDs connected to terminals b and c and number 3 is displayed by lighting the LEDs connected to a, b, c, d and g. Some calculators use seven-segment displays which are LED displays. Other calculators use liquid crystal displays (LCDs), again of the seven-segment display type.

The seven-segment display is a digital output device which changes electrical energy into light energy. Seven-segment displays are used in some televisions, hi-fis, instrument panels and calculators.

Fascinating Physics

Some cars are being fitted with LED clusters instead of filament side and brake lamps. LED clusters contain a number of ultra-bright LED's. The main advantages compared with filament lamps are low power consumption, long life and reliability.

Binary and decimal numbers

Decimal code

Normally we count on the scale of ten, or decimal, system using ten digits 0 to 9. When the count is greater than 9, we place a 1 in the second column to represent tens.

Binary code

Counting in electronic systems is done by digital circuits on the scale of two, or binary system. The digits 0 and 1 are used and are represented by low and high voltages respectively. Many more columns are necessary since the number after 1 in binary is 1 0 (this is the number two in decimal).

Columns from the right in this case represent powers of 2:

$$2^3 = 8 \quad 2^2 = 4 \quad 2^1 = 2 \quad 2^0 = 1$$

Table 4.1 shows how the decimal numbers 0 to 9 are coded in the binary system.

Decimal number	Binary code			
	$2^3 = 8$	$2^2 = 4$	$2^1 = 2$	$2^0 = 1$
0	0	0	0	0
1	0	0	0	1
2	0	0	1	0
3	0	0	1	1
4	0	1	0	0
5	0	1	0	1
6	0	1	1	0
7	0	1	1	1
8	1	0	0	0
9	1	0	0	1

Table 4.1 Coding decimal numbers in binary.

End of Section Questions

1 State the energy conversion for the following output devices:
(a) electric motor; (b) LED; (c) loudspeaker.

2 A student is asked to draw a suitable circuit to light an LED. The circuit must consist of a 6 V battery, a switch, a resistor and the LED.

(a) Draw a circuit diagram which will allow the LED to light when the switch is closed.
(b) Explain why a resistor is required.

3 An LED and a resistor are connected in series to a 9 V battery. The maximum voltage allowed across the LED is 1.8 V and the current through it must not exceed 12 mA. Calculate the value of the resistor required for the circuit.

4 Which LEDs in figure 4.8 on page 110 are required to be lit in order to display the numbers: (a) 2; (b) 5; (c) 7?

5 The following is a list of output devices:

buzzer; lamp; motor; relay; solenoid

Which output device from the list, could be used to:

(a) indicate that the required cooking time for an oven is complete?
(b) move the conveyor belt at a supermarket checkout?
(c) stop and start the movement of the conveyor belt at a supermarket checkout?

6 What binary number represents the decimal number:
(a) 2; (b) 5; (c) 7; (d) 8?

7 What is the decimal number represented by the binary numbers:

(a) 0 0 0 1; (b) 0 1 0 0; (c) 0 1 1 0; (d) 1 0 0 1?

At the end of this section you should be able to:

1 Describe the energy transformations involved in the following devices: microphone; thermocouple; solar cell.
2 State that the resistance of a thermistor changes with temperature and the resistance of an LDR decreases with increasing light intensity.
3 Carry out calculations using $V = IR$ for the thermistor and the LDR.
4 State that during charging the voltage across a capacitor increases with time.
5 Identify from a list an appropriate input device for a given application.

6 Carry out calculations involving voltages and resistances in a voltage divider.
7 State that the time to charge a capacitor depends on the values of the capacitance and the series resistance.
8 Identify appropriate input devices for a given application.

There are a large number of input devices available for different applications.

The microphone

A microphone is connected to an oscilloscope. As louder notes are played into the microphone, the trace on the oscilloscope increases in amplitude.

A **microphone** is an input transducer which changes sound energy into electrical energy. The louder the sound, the greater the electrical energy produced.

The thermocouple

When the junction of the thermocouple, shown in figure 4.9, is placed in a Bunsen flame, the voltmeter reading increases.

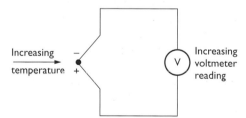

Figure 4.9 A thermocouple connected to a voltmeter.

A **thermocouple** is an input transducer which changes heat energy into electrical energy. The higher the temperature of the junction, the greater the electrical energy produced.

The solar cell

When the solar cell, shown in figure 4.10, is exposed to more light, the voltmeter reading increases.

A **solar cell** is an input transducer which changes light energy into electrical energy. The brighter the light shining on the solar cell, the greater the electrical energy produced.

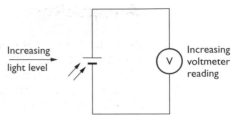

Figure 4.10 A solar cell connected to a voltmeter.

The thermistor

When the thermistor shown in figure 4.11 is heated, the ohmmeter reading decreases.

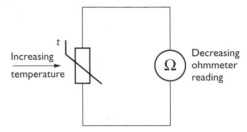

Figure 4.11 A thermistor connected to an ohmmeter.

A **thermistor** is an input device. The resistance of a thermistor decreases when its temperature increases.

Temperature ↑ – resistance of thermistor ↓

The only type of thermistor to be considered in this book will act as shown above, that is the resistance of the thermistor decreases as its temperature increases.

The light-dependent resistor (LDR)

When the light-dependent resistor, shown in figure 4.12, is exposed to more light, the ohmmeter reading decreases.

A **light-dependent resistor** (**LDR**) is an input device. As the light gets brighter (light intensity increases) the resistance of the LDR decreases.

Light intensity ↑ – resistance of LDR ↓

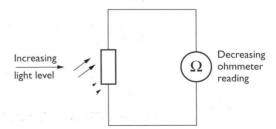

Figure 4.12 A light dependent resistor (LDR) connected to an ohmmeter.

A capacitor

A **capacitor** consists of two metal plates separated by an insulator. The construction of a capacitor is shown in figure 4.13. It is a device which can store electric charge. The units of capacitance are **farads** (F). Most capacitors have very small values and so are measured in microfarads (μF), that is millionths of a farad.

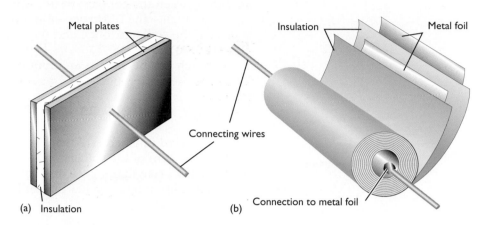

Metal plates — Insulation — Metal foil

Connecting wires

Connection to metal foil

(a) Insulation   (b)

Figure 4.13 (a) A capacitor consists of two metal plates separated by an insulator, (b) practical capacitors are constructed like a 'Swiss roll'. Why is this construction preferred to that shown in (a) for most capacitors?

Figure 4.14(a) shows a capacitor with no charge on its plates. Figure 4.14(b) shows the capacitor when it is fully charged. This occurs when the voltage across the plates of the capacitor is equal to the supply voltage. Figure 4.14(c) shows how a charged capacitor can be discharged by connecting the two plates together using a switch. When the switch is closed, the capacitor is discharged.

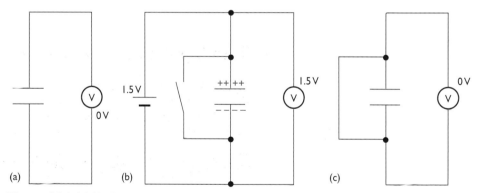

Figure 4.14 (a) Uncharged capacitor, (b) fully charged capacitor, (c) discharged capacitor.

A capacitor can be used as an input device. A discharged capacitor has no voltage across its plates. When a capacitor is charging up, the voltage across the plates takes time to rise to the supply voltage. The voltage across a fully charged capacitor is equal to the supply voltage.

A potentiometer

Figure 4.15 shows an ohmmeter connected to the slider (sliding contact) and one end of a potentiometer. As the slider Z of the potentiometer is moved from X to Y, the ohmmeter reading increases.

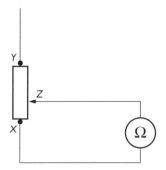

Figure 4.15 An ohmmeter connected to the slider and one end of a potentiometer.

A **potentiometer** can be used as an input device. As the slider Z is moved from X to Y, the resistance of XZ increases while the resistance of ZY decreases.

A switch

An ohmmeter is connected to a switch. When the switch is open, there is an air gap between the contacts of the switch and the ohmmeter reading is very, very high – the resistance of an open switch is infinite. When the switch is closed, the contacts touch and the ohmmeter reading is zero (or very close to zero).

A **switch** can be used as an input device.

Voltage dividers as input devices

You discovered in Chapter 2 that, for a series circuit, the supply voltage was equal to the sum of the voltages across the individual resistors, that is $V_S = V_1 + V_2 + V_3$ (see figure 2.21 on page 43). This means that the supply voltage is split up into smaller bits and this is the basis for the input to an electronic system.

A voltage divider consists of two devices, usually resistors, connected in series as shown in figure 4.16. The supply voltage is divided up into two smaller voltages.

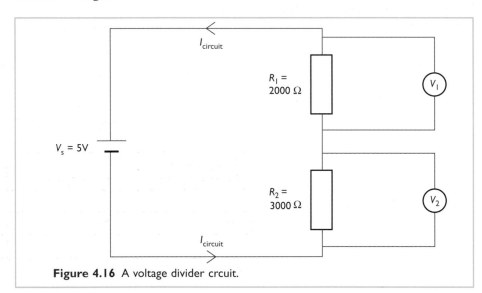

Figure 4.16 A voltage divider crcuit.

$R_T = R_1 + R_2$

$R_T = 2000 + 3000 = 5000\,\Omega$

$I_{\text{circuit}} = \dfrac{V_S}{R_T} = \dfrac{5}{5000} = 0.001\,A$

$V_1 = I_{\text{circuit}}\,R_1 = 0.001 \times 2000 = 2\,V$

$V_2 = I_{\text{circuit}}\,R_2 = 0.001 \times 3000 = 3\,V$

$\dfrac{V_1}{V_2} = \dfrac{2}{3}$ and $\dfrac{R_1}{R_2} = \dfrac{2000}{3000} = \dfrac{2}{3}$

i.e. $\dfrac{V_1}{V_2} = \dfrac{R_1}{R_2}$

The values of V_1 and V_2 depend on the values of R_1 and R_2.

For a voltage divider: $\dfrac{V_1}{V_2} = \dfrac{R_1}{R_2}$

Either voltage V_1 or V_2 could be used as the input voltage to an electronic system, but it is usual to use V_2 as the input voltage to an electronic process device. We shall be interested in what happens to this voltage when changes are made in a voltage divider circuit.

Voltage divider with a thermistor

A graph of resistance against temperature for a thermistor is given in figure 4.17.

Figure 4.18 shows the thermistor connected in series with a $10\,\text{k}\Omega$ resistor to form a voltage divider circuit. The voltages across the thermistor (V_1) and across the resistor (V_2) at different temperatures are shown in table 4.2. (You should check these values by calculation yourself.)

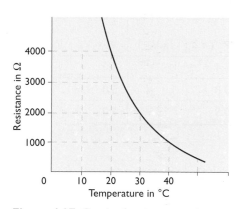

Figure 4.17 Graph of resistance against temperature for a thermistor

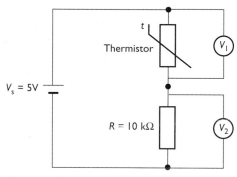

Figure 4.18 Voltage divider circuit for the results shown in table 4.2.

Temperature of thermistor in °C	V_1 (= V_{th}) in V	V_2 (= V_R) in V	$V_s = V_1 + V_2$ in V
20	1.4	3.6	5.0
30	0.8	4.2	5.0
40	0.5	4.5	5.0

Table 4.2

For this circuit, the resistance of the thermistor decreases as its temperature increases and so the voltage across the thermistor (V_1) decreases and hence the voltage across the resistor R (V_2) increases even though the **value of R has not been changed.**

The circuit in figure 4.19 shows the same thermistor and resistor connected in series but with their positions interchanged.

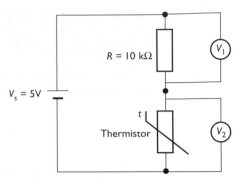

Figure 4.19 Alternative voltage divider circuit. The results for this circuit are shown in table 4.3.

Temperature of thermistor in °C	V_1 (= V_R) in V	V_2 (= V_{th}) in V	$V_s = V_1 + V_2$ in V
20	3.6	1.4	5.0
30	4.2	0.8	5.0
40	4.5	0.5	5.0

Table 4.3

For this circuit, the resistance of the thermistor decreases as its temperature increases and so the voltage across the thermistor (V_2) decreases (and hence the voltage across the resistor increases). See table 4.3.

It should be noted that these two circuits, although containing the same components, give a voltage V_2 which changes in opposite directions – when the temperature increases in the first circuit, V_2 increases, while in the second, V_2 decreases.

Voltage divider with a light-dependent resistor

A graph of resistance for a light-dependent resistor (LDR) against light intensity is given in figure 4.20.

Figure 4.21 shows the LDR connected in series with a 1 kΩ resistor to form a voltage divider circuit. The voltages across the LDR (V_1) and across the resistor (V_2) at different light levels are shown in table 4.4. (You should check these values by calculation yourself.)

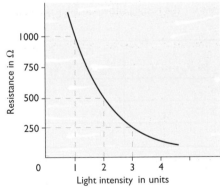

Figure 4.20 Graph of resistance against light intensity for an LDR

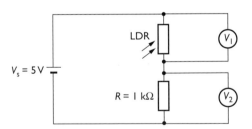

Figure 4.21 Voltage divider circuit for the results shown in table 4.4.

Light intensity in units	V_1 (= V_{LDR}) in V	V_2 (= V_R) in V	$V_s = V_1 + V_2$ in V
1	2.5	2.5	5.0
2	1.7	3.3	5.0
3	1.0	4.0	5.0

Table 4.4

For this circuit, the resistance of the LDR decreases as more light falls on it and so the voltage across the LDR (V_1) decreases and the voltage across the resistor R (V_2) increases, even though the **value of R has not been changed**.

The circuit in figure 4.22 shows the same LDR and resistor connected in series but with their positions interchanged. See table 4.5.

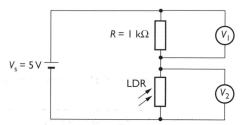

Figure 4.22 Alternative voltage divider circuit. The results for this circuit are shown in table 4.5.

Light intensity in units	V_1 (= V_R) in V	V_2 (= V_{LDR}) in V	V_s = V_1 + V_2 in V
1	2.5	2.5	5.0
2	3.3	1.7	5.0
3	4.0	1.0	5.0

Table 4.5

For this circuit, the resistance of the LDR decreases as more light falls on it and so the voltage across the LDR (V_2) decreases (and so the voltage across the resistor R (V_1) increases).

Voltage divider with a capacitor

The circuit shown in figure 4.23 was used to time how long it took a capacitor to charge up to a certain voltage. The results for different values of capacitor and resistor are shown in table 4.6.

The capacitor was discharged by closing switch S. As soon as the switch was opened, the capacitor began to charge up.

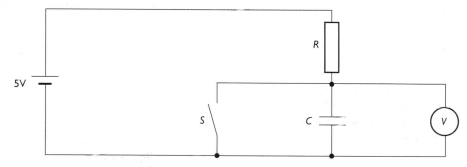

Figure 4.23 Voltage divider circuit for a capacitor and a resistor.

Value of C in μF	Value of R in kΩ	Time taken for C to charge up in s
1000	10	60
1000	1	6
200	10	12

Table 4.6

For the above circuit, the voltage across the capacitor (V) increases to the supply voltage, but the time taken is dependent on the values of C and R:

♦ Increasing the value of the capacitor increases the time taken to charge it up, i.e. a larger capacitor is able to store more charge.

> Capacitance ↑ – time taken ↑

♦ Increasing the value of the series resistor decreases the charging current and so fewer charges flow onto the capacitor plates in one second. The capacitor therefore takes longer to become fully charged.

> Resistance ↑ – time taken ↑

Capacitor charges up – voltage across it increases (to the supply voltage).
Capacitor discharges – voltage across it decreases (to zero).

Voltage divider with a potentiometer

Figure 4.24 shows a voltmeter connected to the slider and one end of a potentiometer. The voltmeter reading increases as the slider Z moves from X to Y of the potentiometer.

As the slider Z is moved from X to Y, the resistance of XZ increases, while the resistance of ZY decreases. The voltage across XZ therefore increases, i.e. the output voltage V (V_{XZ}) increases (while the voltage across ZY decreases).

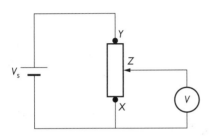

Figure 4.24 Voltage divider circuit for a potentiometer

Voltage divider with a switch

Figures 4.25(a) and (b) show a switch connected in series with a resistor to form a voltage divider circuit.

In circuit (a), when the switch S is open $V = 0\,V$, and when S is closed $V = 5\,V$.

In circuit (b), when the switch S is open $V = 5\,V$, and when S is closed $V = 0\,V$.

The easiest way to understand these two circuits is to consider the switch being open, i.e. when there is no current. When the current through a resistor is zero then the voltage across it is zero. Since the switch and the resistor form a voltage divider circuit and $V_S = V_1 + V_2$, then the voltage across the switch must equal V_S, i.e. 5 V.

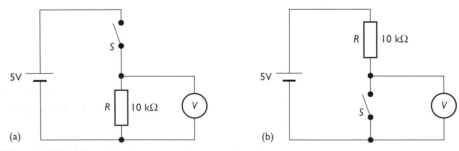

Figure 4.25 Voltage divider circuits for a switch and a resistor.

Section 4.3 Summary

◆ Most input devices change some form of energy into electrical energy.
◆ There are a number of input devices, examples are: microphone; thermocouple; solar cell; thermistor; light dependent resistor (LDR); capacitor; potentiometer; switch.
◆ The resistance of a thermistor changes with temperature – for the thermistors considered in this book the resistance of a thermistor decreases as the temperature increases.
◆ The resistance of an LDR decreases with increasing light level.
◆ An uncharged capacitor has no voltage across it.
◆ As a capacitor charges up, the voltage across the capacitor increases with time until it reaches the same value as the supply voltage.
◆ The time taken to charge a capacitor depends on the values of the capacitance and the series resistor – increasing the value of the capacitor increases the time taken or increasing the value of the series resistor increases the time taken.

◆ For the following circuit: $\dfrac{V_1}{V_2} = \dfrac{R_1}{R_2}$

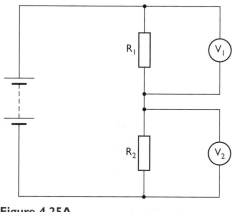

Figure 4.25A

End of Section Questions

1 What is the energy conversion for:

(a) a microphone? **(b)** a thermocouple? **(c)** a solar cell?

2 Name an appropriate input device which could be used as part of a circuit for the following:

(a) an electronic temperature sensor;
(b) an electronic light meter;
(c) an electronic timer;
(d) an electronic sound sensor.

3 The diagram shows two resistors connected to a 6 V battery.

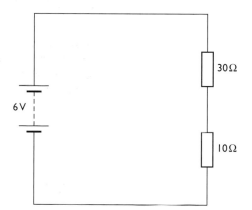

Figure 4.3Q3

(a) What is the total resistance of the circuit?
(b) Find the current drawn from the battery.
(c) Calculate the voltage across the 10 Ω resistor.

4 In the following circuits calculate the voltages V_1 and V_2.

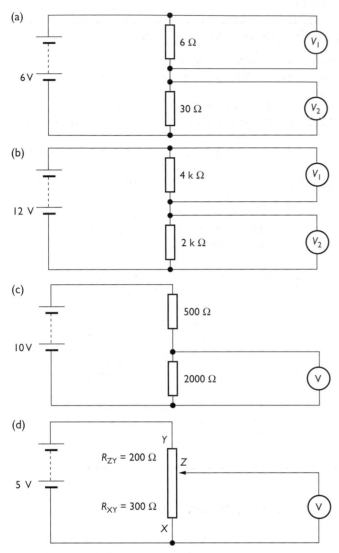

Figure 4.3Q4

5 A thermistor is connected in the circuit shown in the diagram. The reading on the ammeter is 0.0036 A and the reading on the voltmeter is 1.8 V when the thermistor is at a temperature of 18°C.

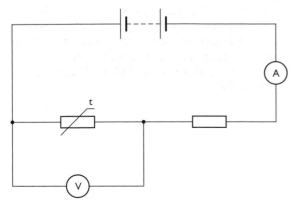

Figure 4.3Q5

(a) Calculate the resistance of the thermistor.
(b) The temperature of the thermistor rises to 20°C. Suggest a suitable reading for the ammeter.

6 The light intensity falling on the LDR in the diagram increases. Describe and explain how the voltmeter reading changes when the light intensity increases.

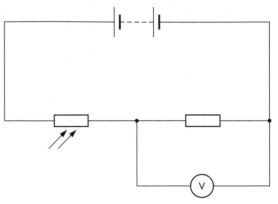

Figure 4.3Q6

7 The capacitor shown in the figure charges when the switch is closed.

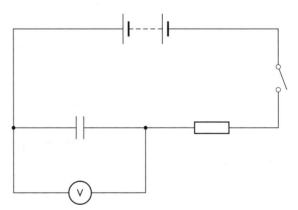

Figure 4.3Q7

(a) What happens to the reading on the voltmeter as the capacitor charges?
(b) The capacitor takes 30 s to charge to 5.0 V. Suggest how the circuit could be altered to give a longer time of charge.

At the end of this section you should be able to:

1 State that a transistor can be used as a switch.
2 State that a transistor may be conducting or non-conducting, i.e. ON or OFF.
3 Draw and identify the circuit symbol for an NPN transistor.
4 Identify from a circuit diagram the purpose of a simple transistor switching circuit.
5 Draw and identify the symbols for two input AND and OR gates, and a NOT gate.
6 State that logic gates may have one or more inputs and that a truth table shows the output for all possible input combinations.
7 State that high voltage = logic '1'; low voltage = logic '0'.
8 Draw the truth tables for two input AND- and OR-gates, and a NOT-gate.
9 Explain how to use combinations of digital logic gates for control in simple situations.
10 State that a digital circuit can produce a series of clock pulses.
11 Give an example of a device containing a counter circuit.
12 State that there are circuits which can count digital pulses.
13 State that the output of the counter circuit is in binary.
14 State that the output of a binary counter can be converted to decimal.

15 Explain the operation of a simple transistor switching circuit.
16 Identify the following gates from truth tables: two-input AND-; two-input OR-; NOT- (inverter).
17 Complete a truth table for a simple combinational logic circuit.
18 Explain how a simple oscillator built from a resistor, capacitor and inverter operates.
19 Describe how to change the frequency of the clock.

A process device – the transistor

We have looked at how signals may be passed into an electronic system by the input part of the system. We have looked at how the desired output may be obtained. It is time to look at how the input signal is modified by the process part of the electronic system.

A transistor has three terminals called the **base**, the **emitter** and the **collector**. The symbol for an NPN transistor is shown in figure 4.26. A transistor can be considered as an electronic switch with no moving parts. The switching is controlled by the voltage applied to the emitter-base. The transistor is off (non-conducting) when the emitter-base voltage is below a certain value, i.e. the electronic switch is open. For the type of transistor we are using this is about 0.7 V, although this voltage varies from one type of transistor to another. However, the transistor is on (conducting) when the emitter-base voltage is equal to or above this certain value (≥ 0.7 V), i.e. the electronic switch is closed (figure 4.27).

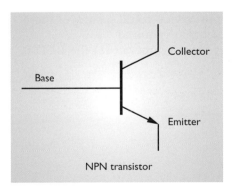

Figure 4.26 Circuit symbol for a transistor.

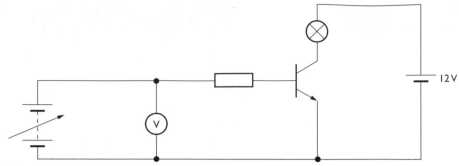

Figure 4.27 The lamp lights when the voltmeter reading is greater than or equal to 0.7 V. When it is less than 0.7 V the lamp does not light.

Figure 4.28 shows an electronic system, i.e. an input (potentiometer), process (transistor) and an output (LED). The voltage across XZ (V_{XZ}) is the input voltage to the transistor. The voltage across XZ (V_{XZ}) has to reach a certain value (≥ 0.7 V) before the transistor will switch on and the output LED light.

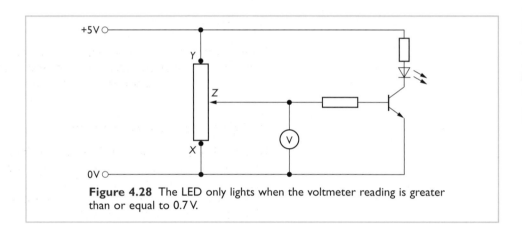

Figure 4.28 The LED only lights when the voltmeter reading is greater than or equal to 0.7 V.

A temperature-controlled circuit

Figures 4.29(a) and 4.29(b) show two temperature-controlled circuits in an electronic system, i.e. an input (voltage divider with a thermistor), process (transistor) and an output (LED). The output device could be any of these discussed in Section 4.2, depending on the output required.

The variable resistor is adjusted until (at room temperature) the LED is just off (see figures 4.29(a) and (b)).

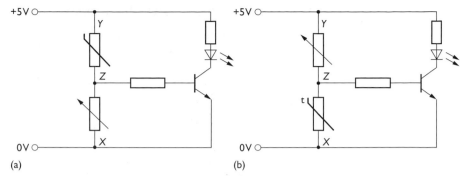

(a)   (b)

Figure 4.29 (a) Temperature-controlled circuit, (b) alternative temperature-controlled circuit.

LED is off
Heat thermistor

$R_{\text{thermistor}}\downarrow$

$V_{\text{thermistor}}\downarrow$ so $V_{\text{ZY}}\downarrow$

$V_{\text{XZ}}\uparrow$

Transistor switches on

LED lights

LED is off
Cool thermistor

$R_{\text{thermistor}}\uparrow$

$V_{\text{XZ}}\uparrow$

Transistor switches on

LED lights

Note: A variable resistor is used in this type of circuit instead of a fixed resistor. The variable resistor allows the circuit to be adjusted to different conditions (temperature in this case) before the output device comes on (or goes off).

A light-controlled circuit

Figures 4.30(a) and 4.30(b) show two light-controlled circuits in an electronic system, i.e. an input (voltage divider with a LDR), process (transistor) and an output (LED).

The variable resistor is adjusted until at normal light level the LED is just off (see figures 4.30(a) and(b)).

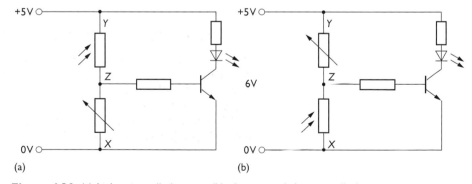

(a)   (b)

Figure 4.30 (a) Light-controlled circuit, (b) alternative light-controlled circuit.

LED is off
Shine more light on LDR

$R_{\text{LDR}}\downarrow$

$V_{\text{LDR}}\downarrow$ so $V_{\text{ZY}}\downarrow$

$V_{\text{XZ}}\uparrow$

Transistor switches on

LED lights

LED is off
Cover LDR

$R_{\text{LDR}}\uparrow$

$V_{\text{XZ}}\uparrow$

Transistor switches on

LED lights

A time-controlled circuit

Figures 4.31(a) and 4.31(b) show two time-controlled circuits in an electronic system, i.e. an input (voltage divider with a capacitor), process (transistor) and an output (LED).

The switch S is closed to discharge the capacitor (voltage across capacitor, $V_C = 0$ V when discharged), then opened to allow the capacitor to start charging.

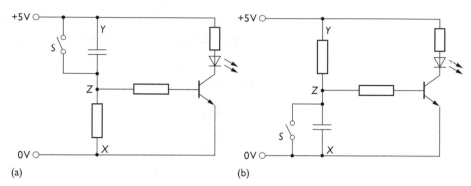

Figure 4.31 (a) Time-controlled circuit, (b) alternative time-controlled circuit.

Switch S closed
$V_{capacitor} = 0$ V

$V_{ZY} = 0$ V

$V_{XZ} = 5$ V

Transistor switches on
LED lights
Open switch S

$V_{capacitor}\uparrow$ so $V_{ZY}\uparrow$

$V_{XZ}\downarrow$

Transistor switches off after
   a short delay
LED goes out

Switch S closed
$V_{capacitor} = 0$ V

$V_{XZ} = 0$ V

Transistor switches off

LED is off
Open switch S
$V_{capacitor}\uparrow$

$V_{XZ}\uparrow$

Transistor switches on after
   a short delay
LED lights

A switch-controlled circuit

Figures 4.32(a) and 4.32(b) show two switch-operated circuits in an electronic system, i.e. an input (voltage divider with a switch), process (transistor) and an output (LED).

When a switch is open in either circuit, no current passes through the resistor and so the voltage across the resistor, $V_{resistor}$, is zero and the voltage across the switch, V_{switch}, is 5 V.

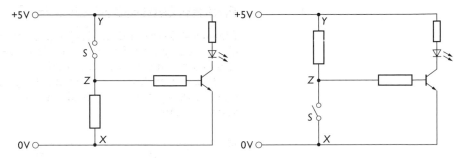

Figure 4.32 (a) Switch-controlled circuit, (b) alternative switch-controlled circuit.

Switch S open
$V_{ZY} = 5\,V$
$V_{XZ} = 0\,V$
Transistor switches off
LED is off
Close switch S
$V_{ZY} = 0\,V$
$V_{XZ} = 5\,V$
Transistor switches on
LED lights

Switch S open
$V_{XZ} = 5\,V$
Transistor switches on
LED lights
Close switch S
$V_{XZ} = 0\,V$
Transistor switches off
LED goes out

Other process devices – logic gates

Logic gates

These are manufactured on a tiny, single chip, called an **integrated circuit (IC)**.

Logic gates are digital devices which frequently obtain analogue signals as their inputs. These analogue inputs have to be converted into a digital form by the gate before it is able to carry out its task.

The physical inputs and outputs from a gate are voltages which may either be 'high' (close to the supply voltage) or 'low' (near to zero volts). These are referred to as logic '1' and logic '0'.

A table known as a **truth table** shows how the output of the gate varies with the input or inputs. A truth table is a shorthand way to show the behaviour of an electronic system.

The Not-(or inverter) gate

Figure 4.33 shows the circuit symbol and truth table for a **NOT-gate**.

From the truth table it can be seen that the output of a NOT-gate is 'not' (the same as) the input. A NOT-gate is also called an **inverter**.

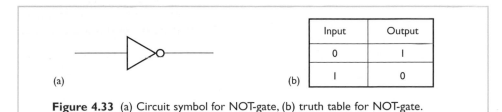

(a)

(b)

Input	Output
0	1
1	0

Figure 4.33 (a) Circuit symbol for NOT-gate, (b) truth table for NOT-gate.

The AND-gate

The circuit symbol and truth table for an **AND-gate** are shown in figure 4.34.

From the truth table, it can be seen that the output of an AND-gate will be logic '1' (high) only, when inputs A and B are logic '1' (high).

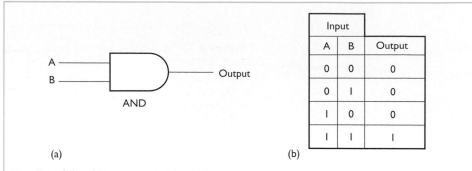

(a)   (b)

Figure 4.34 (a) Circuit symbol for AND-gate, (b) truth table for AND-gate.

Input		Output
A	B	
0	0	0
0	I	0
I	0	0
I	I	I

The OR-gate

The circuit symbol and truth table for an **OR-gate** are shown in figure 4.35.

From the truth table, it can be seen that the output of an OR-gate will be logic '1' (high) when either of the inputs A or B is '1' (high).

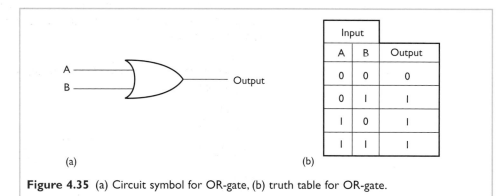

(a)   (b)

Figure 4.35 (a) Circuit symbol for OR-gate, (b) truth table for OR-gate.

Input		Output
A	B	
0	0	0
0	I	I
I	0	I
I	I	I

The AND-, OR- and NOT-gates above may be combined together to form a sophisticated electronic system.

In the following examples you may assume that a light sensor gives out a logic '1' in light and a logic '0' in dark; and a temperature sensor gives out a logic '1' when warm and a logic '0' when cold.

Example 1
Draw a logic diagram and truth table for a warning LED to light when a car engine gets too hot. The lamp should only operate when the ignition of the car is switched on (logic '1').

Solution
Require LED to be on (1) when ignition is on (1) and engine is too hot (1).

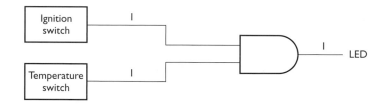

Temperature sensor	Ignition switch	LED
cold (0)	off (0)	off (0)
cold (0)	on (1)	off (0)
warm (1)	off (0)	off (0)
warm (1)	on (1)	on (1)

Figure 4.36 Logic diagram and truth table for example 1.

Example 2
Draw a logic diagram and truth table which will switch on the pump of a central heating system when the house is cold and the central heating is switched on (logic '1').

Solution
Require pump on (1) when central heating is on (1) and temperature is cold (0) (not warm (1)).

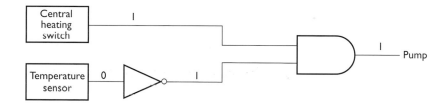

Temperature sensor	Central heating switch	Pump
cold (0)	off (0)	off (0)
cold (0)	on (1)	on (1)
warm (1)	off (0)	off (0)
warm (1)	on (1)	off (0)

Figure 4.37 Logic diagram and truth table for example 2.

Example 3

Draw a logic diagram and truth table which will turn on a heater in a greenhouse when it gets cold at night. The heater should be switched off (logic '0') during the day.

Solution

Require heater on (1) when it is cold (0) (not warm (1)) and dark (0) (not light (1)).

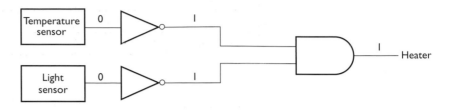

Light sensor	Temperature sensor	Heater
dark (0)	cold (0)	on (1)
dark (0)	warm (1)	off (0)
light (1)	cold (0)	off (0)
light (1)	warm (1)	off (0)

Figure 4.38 Logic diagram and truth table for example 3.

The clock-pulse generator

Clock pulses are pulses of voltage which occur regularly, like the ticking of a clock. The duration of the pulses and the rate at which pulses occur can be varied by altering the values of the capacitor or resistor to suit different applications. Clock pulses are mainly used in counting and timing signals, e.g. digital watches, computers and in the operation of traffic lights.
The circuit shown in figure 4.39 can be used in counting and timing.

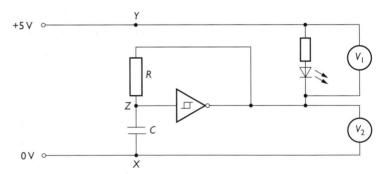

Figure 4.39 A clock-pulse generator circuit.

Note:
◆ The input to the NOT-gate (inverter) is the voltage across the capacitor (V_{XZ}).
◆ The supply voltage of 5 V is equal to $V_1 + V_2$. When the output of the NOT-gate is logic '0', voltmeter $V_2 = 0$ V and so voltmeter $V_1 = 5$ V. Since there are 5 volts across the LED and resistor, the LED lights. When the output of the NOT-gate is logic '1', voltmeter $V_2 = 5$ V and so $V_1 = 0$ V. Since there is no voltage across the LED and resistor, the LED does not light.

A simple explanation of how this circuit operates is shown in table 4.7.

Capacitor	Input to NOT-gate	Output from NOT-gate	V_2 in V	V_1 in V	LED
Charged	I	0	0	5	Lit
Discharged	0	I	5	0	Unlit
Charged	I	0	0	5	Lit
Discharged	0	I	5	0	Unlit

Table 4.7

The pattern in table 4.7 continues so that the generator produces pulses. If the value of the capacitor is increased, it takes longer to charge and discharge, so there are fewer pulses per second, i.e. their frequency is lower. If the value of the resistor is increased, it takes longer for the capacitor to charge and discharge – fewer pulses per second – lower frequency. This means that the values of R and C control the frequency of the pulses.

Counting circuits

A counter is an electronic circuit which is able to count the electrical pulses of the clock-pulse generator. Figure 4.40 shows a counter connected to a timing circuit.

Table 4.8 shows the number of pulses from the timing circuit and the display on the counter board (a logic '1' is equal to a lit LED).

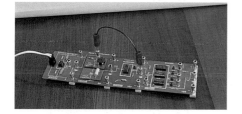

Figure 4.40 An electronic counter circuit. The circuit counts in binary.

Number of clock pulses	(8) D	(4) C	(2) B	(1) A
0	0	0	0	0
I	0	0	0	I
2	0	0	I	0
3	0	0	I	I
4	0	I	0	0
5	0	I	0	I
6	0	I	I	0
7	0	I	I	I
8	I	0	0	0
9	I	0	0	I

Table 4.8

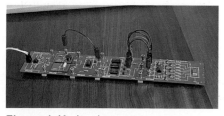

Figure 4.41 An electronic counter circuit. The circuit counts in both binary and decimal.

The completed table shows the binary code for the decimal numbers from zero to nine. A circuit called a **binary-to-decimal decoder** together with a seven-segment display is used to convert the binary display on the counter to a decimal number on the seven-segment display (figure 4.41).

- A transistor is an electrically operated switch. The circuit symbol for a NPN transistor is shown in figure 4.26.
- A transistor is non-conducting (OFF) for voltages below a certain value (normally below 0.7 V) but conducting (ON) at voltages at or above this certain value (at or above 0.7 V).
- Logic gates are digital devices – NOT (or Inverter), AND and OR gates.
- Logic gates use logic '1' to represent a high voltage level and logic '0' to represent a low voltage level.
- A truth table shows how the output from a gate or system varies with the input or inputs.
- The circuit symbols and truth tables for NOT, AND and OR gates are shown in figures 4.33, 4.34 and 4.35.
- A timing circuit or oscillator can be built from a resistor, capacitor and a NOT gate (Inverter) to give clock pulses. See figure 4.39.
- The clock pulses can be made more frequent (frequency increased) by decreasing the value of the capacitor or by decreasing the value of the resistor.
- The clock pulses (zeros and ones) are digital and can be counted using a counter circuit.
- The output of a counter is in binary and this can be changed into a decimal number using a binary to decimal circuit.

End of Section Questions

1 The diagram shows an electronic circuit.

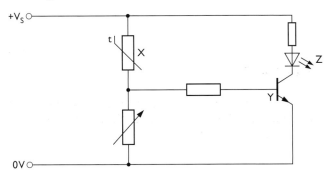

Figure 4.4Q1

(a) Name the components X, Y and Z in the circuit.
(b) What is the purpose of component Y in the circuit?

2 Draw the circuit symbol and truth table for the following logic gates:

(a) NOT; (b) AND; (c) OR

3 Copy and complete the truth table for the circuit shown in the figure below.

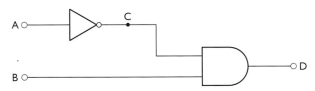

Figure 4.4Q3

A	B	C	D
0	0		
0	1		
1	0		
1	1		

4 Copy and complete the truth table for the circuit shown in the figure below.

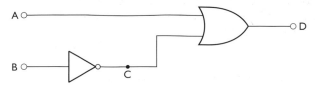

Figure 4.4Q4

A	B	C	D
0	0		
0	1		
1	0		
1	1		

5 A student is asked to design a circuit to remind a car driver to fasten his or her safety belt. A buzzer is to sound when the driver's seat belt remains unclipped but only when the ignition is switched on.
The safety belt sensor gives logic '1' when fastened.
The ignition switch gives logic '1' when switched on.
The buzzer is switched on by logic '1'.
Draw a suitable logic diagram, naming the logic gates involved.

6 The diagram shown in the figure shows a circuit to detect the light level in a room.

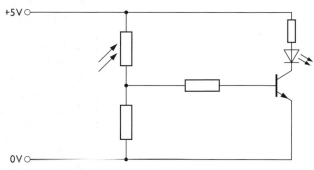

Figure 4.4Q6

The room is initially dark. Describe and explain what will happen as the light level in the room increases.

7 The circuit shown in the figure shows a clock pulse generator. The generator has a frequency of 10 Hz.

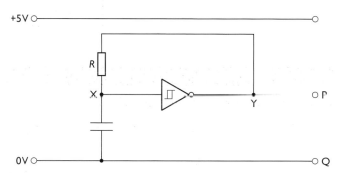

Figure 4.4Q7

The capacitor in the circuit is replaced with a capacitor of a higher value. What effect does this have on the frequency of the pulses?

At the end of this section you should be able to:

1 Identify from a list, devices in which amplifiers play an important part.
2 State the function of the amplifier in devices such as radios, intercoms and music centres.
3 State that the output signal of an audio amplifier has the same frequency as, but larger amplitude than, the input signal.
4 Carry out calculations involving input voltage, output voltage and voltage gain of an amplifier.

5 Describe how to measure the voltage gain of an amplifier.
6 State that power may be calculated from V^2/R where V is the voltage and R the resistance (impedance) of the circuit.
7 State that the power gain of an amplifier is the ratio of power output to power input.
8 Carry out calculations involving the power gain of an amplifier.

An amplifier is an analogue process device which is generally used to make electrical signals larger. For instance, a CD player will give a signal of perhaps 10 mV (0.010 V). You would not be able to hear this tiny signal from the CD player if it was directly connected to a loudspeaker. The amplitude of this signal has to be made bigger, i.e. amplified before it can be connected to the loudspeaker.

Figure 4.42(a) shows an amplifier being used to make the electrical signal from a signal generator larger. The input signal to and output signal from an amplifier are connected to two identical oscilloscopes. The traces displayed on the oscilloscope screens are shown in figure 4.42(b).

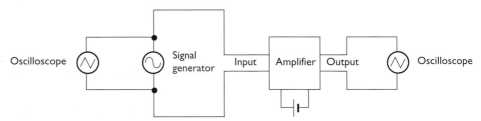

Figure 4.42 (a) This amplifier circuit makes the output voltage larger than the input voltage.

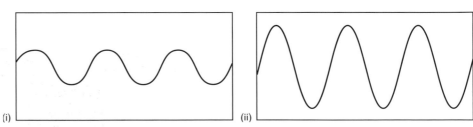

Figure 4.42 (b) (i) Input voltage, (ii) output voltage.

The amplitude of the output signal from an amplifier is larger than the amplitude of the input signal – the extra energy comes from the electrical supply to the amplifier. However, the frequency of the output signal from the amplifier is the same as the frequency of the input signal. In audio appliances the amplifier is generally the volume control.

Amplifiers

Amplifiers are to be found in many devices, e.g. radios, televisions, hi-fis, intercoms and loudhailers. Two of these are shown in figure 4.43.

Voltage gain

If an amplifier has an input voltage of 0.5 V and an output voltage of 5.0 V, then the output voltage has increased by 10 times, i.e. the voltage gain is 10:

$$\text{Voltage gain} = \frac{\text{output voltage}}{\text{input voltage}}$$

Note that voltage gain does not have a unit.

In most amplifiers, the voltage gain is from a series of transistors. The output from one is fed into the input of the next and so on. Usually the transistors form part of an integrated circuit (IC).

Example
Figure 4.44 shows the apparatus used to measure the voltage gain of an amplifier.
(a) State the input and output voltages to/from the amplifier.
(b) Calculate the voltage gain of the amplifier.

Solution

(a)   Input voltage = 0.1 V; output voltage = 1.5 V

(b) $\text{Voltage gain} = \dfrac{\text{output voltage}}{\text{input voltage}} = \dfrac{1.5}{0.1} = 15.$

Figure 4.43 (a) Input: Small a.c. voltage from microphone. Output: larger a.c. voltage across loudspeaker. (b) Input: small a.c. voltage from aerial. Output: larger a.c. voltage across loudspeaker.

Power gain of amplifiers

The voltage gain is not particularly useful when comparing amplifiers, for example, in different hi-fi systems. It is better to consider the **power gain** of the amplifier:

$$\text{power gain} = \frac{\text{power output}}{\text{power input}}$$

Like voltage gain, power gain does not have a unit. Power amplifiers have large power gains. Since most amplifiers have roughly the same power input, it is only necessary to give the maximum power output to be able to compare amplifiers.

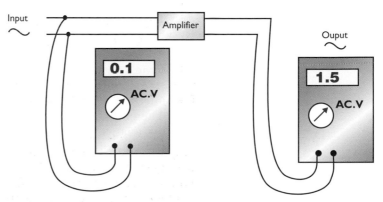

Figure 4.44 The voltage gain of the amplifier can be calculated using the input and output voltages shown

$$\text{Power} = \text{current} \times \text{voltage} = IV \qquad \text{but } V = IR \text{ (Ohm's law)}$$

$$\text{Power} = IV = I \times (IR) = I^2 R$$

$$\text{Power} = IV = V \times \frac{V}{R} = \frac{V^2}{R} \text{ as } I = \frac{V}{R}$$

i.e. $$\text{Power} = IV = I^2 R = \frac{V^2}{R}$$

$\text{Power} = \dfrac{V^2}{R}$ **is a useful formula for finding either the input or output power of an amplifier.**

Example

A girl connects a set of headphones of resistance $16\,\Omega$ to her personal stereo. The amplifier in the stereo produces $0.04\,\text{W}$ of power in the headphones.
(a) What is the voltage applied to the headphones?
(b) Calculate the input power to the stereo amplifier when the power gain is 20.

Solution

(a)
$$P = \frac{V^2}{R}$$

$$0.04 = \frac{V^2}{16}$$

$$V^2 = 0.04 \times 16 = 0.64$$

$$V = 0.8\,\text{V}$$

(b)
$$\text{Power gain} = \frac{\text{output power}}{\text{input power}}$$

$$20 = \frac{0.04}{\text{input power}}$$

$$\text{input power} = \frac{0.04}{20} = 0.002\,\text{W} = 2\,\text{mW}$$

Artificial limbs

Fascinating Physics

Twenty years ago, artificial limbs for disabled people had little control and provided limited mobility. However, the advance of electronic systems together with new materials and technologies in mechanical engineering has enabled the artificial limb of today to provide greater increased mobility and finer control of movements. Artificial limbs, such as fingers, hands and legs, are available to improve the quality of life of a disabled person.

Section 4.5 Summary

◆ An amplifier is an analogue process device which makes electrical signals larger.
◆ Audio amplifiers are found in devices such as radios, televisions, hi-fi's, intercoms and loudhailers.
◆ The output signal from an audio amplifier has the same frequency as, but a larger amplitude than, the input signal.
◆ For an amplifier:

$$\text{Voltage gain} = \frac{\text{output voltage}}{\text{input voltage}} \qquad \text{and Power gain} = \frac{\text{output power}}{\text{input power}}$$

◆ Power, $P = IV = I^2 R = \dfrac{V^2}{R}$

1   A number of household electrical devices are listed below.

Computer, food mixer, hi-fi, iron, microwave oven, radio, table lamp, television, vacuum cleaner.

From the above list write down the names of the devices in which amplifiers play an important part.

2   An amplifier is a major component in a loudhailer.

   (a) What is the purpose of the amplifier in a loudhailer?
   (b) How does the frequency of the input signal compare with the output signal from an amplifier?

3   The input signal to an audio amplifier is 3.0 mV. The output signal is 1.5 V. Calculate the voltage gain of this amplifier.

4   The input voltage to an amplifier is 0.022 V. The voltage gain of the amplifier is 250. Calculate the voltage at the output from the amplifier.

5   The power gain of an amplifier is 360. Calculate the power at the input terminals of the amplifier when the power at its output terminals is 12 W.

6   The input voltage to an amplifier is 14 mV. The input resistance of the amplifier is 20 kΩ.

   (a) Calculate the power input to the amplifier.
   (b) The power output of the amplifier is 15 W. Calculate the power gain of the amplifier.

I Two students are asked to design an electronic timer and an electronic thermometer. They are supplied with the following information about some electronic devices.

Capacitor	220 μF
Light emitting diode	Standard 5 mm diameter; max current 25 mA; max voltage 2.5 V
Loudspeaker	38 mm diameter, 8 Ω
Motor	2300 revolutions at 1.5 V, 0.26 A
NOT gate	74LS04
Relay	Coil voltage 6 V; single pole changeover switch rated at 10 A, 24 V d.c.
Solar cell	45 × 26 × 7.5 mm
Thermistor	Resistance at 10°C = 3200 Ω; Resistance at 20° C = 1600 Ω; Resistance at 30° C = 1000 Ω
Transistor	BC 548B TO92 case NPN GP transistor

a) Name **two** devices in the list which are process devices.

b) State the useful energy change which takes place in the solar cell.

c) Which of the devices listed above would be a suitable input device for the electronic timer?

d) The students' design for the electronic thermometer circuit is shown in the figure below.

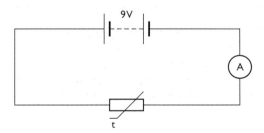

Calculate the current in the circuit when the thermistor is at a temperature of 20° C.

2 A light dependent resistor (LDR) and a 1 kΩ resistor are connected to a 9 V battery as shown in the figure below.

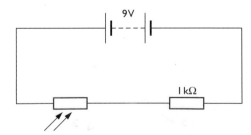

a) When the LDR is in moderate light, the voltage across the LDR is 5 V.
  (i) What is the voltage across the resistor?
  (ii) Calculate the current drawn from the battery.
  (iii) Calculate the resistance of the LDR.

b) The light intensity falling on the LDR now increases.

Explain what happens to the voltage across the resistor.

3 In the circuit shown in the figure below, the variable resistor is adjusted so that the LED is just off when the temperature of the thermistor is 20° C.

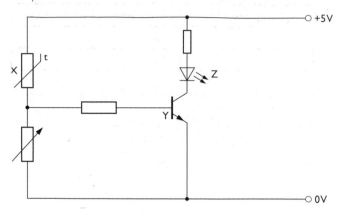

a) Name the components X, Y and Z.

b) The resistance of the thermistor decreases as its temperature increases. Explain what happens when the temperature of the thermistor rises to 22° C.

c) The circuit could be altered to warn a gardener of low temperature conditions. What alteration should be made to the circuit diagram that would allow the LED to light during low temperatures?

4 A student has to design a circuit that will allow an LED to light. The circuit should contain a 6 V battery, a switch, a resistor and the LED.

a) Draw a suitable diagram, which will allow the LED to light when the switch is closed.

b) The voltage across the lit LED is 2.4 V when the current in the LED is 20 mA. The selection of resistors available for the circuit are 120 Ω, 180 Ω, and 300 Ω.

Show by calculation which resistor is required for the circuit.

5 Part of the circuit which operates an automatic light is shown in the figure below.

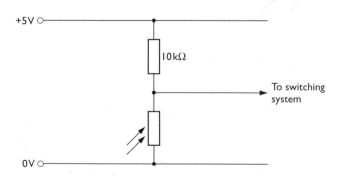

The resistance of the light dependant resistor (LDR) in different lighting conditions is shown in the table below.

Lighting condition	Resistance of LDR in kΩ
Light	0.5
Dark	90

a) The LDR is placed in darkness. Calculate the voltage across the LDR.
b) A lamp rated at 60 W, 230 V is connected to the light sensor using the circuit shown in the figure below.

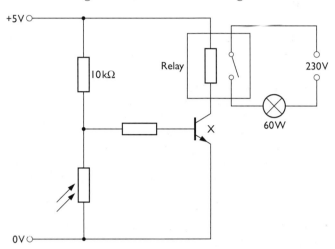

(i) Name component X.
(ii) When the LDR is in darkness, the 60 W lamp lights. Explain how this happens.
(iii) Calculate the current in the 60 W lamp when it is lit.

6 Voltage pulses can be generated using the circuit shown in the figure below.

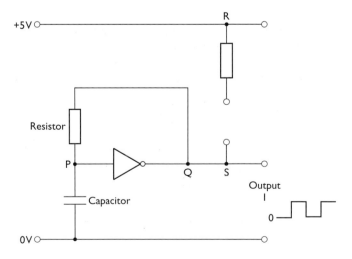

a) Copy and complete the table below to show the logic level at P and Q when the capacitor in the circuit is charged and when it is uncharged.

	Logic level at P	Logic level at Q
Capacitor charged		
Capacitor uncharged		

b) An LED is to be connected in the gap between R and S.
(i) Copy and complete the part of the circuit diagram between R and S. Show an LED correctly connected in the gap between R and S.
(ii) The LED is lit. What is the logic level at Q?
c) The frequency of the voltage pulses is to be reduced. Give **one** change to the circuit which would allow the frequency of the pulses to be reduced.

7 The block diagram for a public address system is shown below.

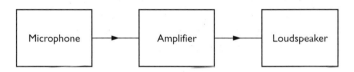

a) The amplifier produces an output power of 2.0 W. The loudspeaker has a resistance of 65 Ω. Calculate the output voltage from the amplifier.
b) The voltage across the input to the amplifier is 0.38 V. Calculate the voltage gain of the amplifier.

CHAPTER FIVE
Transport

Bill was driving home after a hard day's work. It was a wet day and his concentration was not as good as normal. Suddenly the lorry in front put on its brakes. When he saw the brake lights come on Bill slammed on his brakes. He noticed that he took longer to come to a halt due to the wet roads. Bill had left sufficient distance to stop but he had been travelling too fast for the conditions of the road surface. He had been accelerating down that hill and the speed of the car was more than the speed limit. He would need to be more careful in future. Indeed he started to wonder about how speed and acceleration were measured, particularly using speed cameras. What was meant by speed and acceleration? He would ask his daughter Sherry who was studying physics. She could explain these ideas to him!

This chapter looks at the ideas of speed and acceleration and how we measure them. Later in the chapter we use these quantities to calculate other useful aspects such as forces.

At the end of this section you should be able to:

1 Describe how to measure an average speed.
2 Carry out calculations involving the relationship between distance, time and average speed.
3 Describe how to measure instantaneous speeds.
4 Explain the terms speed and acceleration.
5 Calculate acceleration from change of speed per unit time (miles per hour per second or metres per second per second).
6 Draw speed–time graphs showing steady speed, slowing down and speeding up.
7 Describe the motions represented by a speed–time graph.
8 Calculate acceleration, from speed–time graphs, for motion with a single constant acceleration.
9 Identify situations where average and instantaneous speeds are different.
10 Explain how the method used to measure the time of travel can have an effect on the measured value of the instantaneous speed.
11 Calculate distance gone and acceleration from speed–time graphs for motion involving more than one constant acceleration.
12 Carry out calculations involving the relationship between initial speed, final speed, time and uniform acceleration.

Speed

Speed is the distance travelled by an object in one second:

$$\text{Speed} = \frac{\text{distance travelled}}{\text{time taken}}$$

If you travelled a distance of 250 m from A to B in 100 s, your average speed is 2.5 metres per second (2.5 m/s). However, your speed will probably vary throughout the journey, particularly if you divided the journey at a point X, e.g. slower than 2.5 m/s at one stage (up to X) and faster at another stage (past X).

$$\text{Average speed} = \frac{\text{total distance travelled}}{\text{total time taken}} = \frac{d}{t}$$

Speed and average speed have the same units of m/s.

Measuring average speed

Using a stopwatch, trolley and measuring tape

The distance is measured between two points X and Y a few metres apart on the ground. The time for the journey is measured by starting a stopwatch when the trolley reaches X and stopping the watch when it reaches Y (figure 5.1).

Figure 5.1 Measuring speed.

Instantaneous speed and average speed

If we could measure very small time intervals, for example hundredths or even thousands of a second, then we could measure the speed of any object just at that moment in time.

◆ Instantaneous speed is the speed of an object at a particular time (instant).
◆ Average speed is the steady speed an object has, to cover the distance in the time allowed.

$$\text{Average speed} = \frac{\text{total distance travelled}}{\text{total time taken}} = \frac{d}{t}$$

When the time taken (t) is very small, the closer the average speed is to the instantaneous (actual) speed.

Example
A body moves a distance of 80 m in a time of 20 s.

$$\text{Average speed between A and B} = \frac{80}{20} = 4\,\text{m/s}$$

The time interval between the beginning and end of the complete distance AB is too large to give a reasonable estimate of the instantaneous speed at X. But the time interval between C and D is much smaller, so gives a much better estimate of the (actual) speed at X. This is shown in figure 5.2.

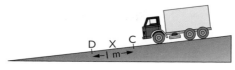

Figure 5.2

Distance CD = 1.1 m

Time to travel from C to D = 0.25 s

Average speed between C and D = $\dfrac{1.1}{0.25}$ = 4.4 m/s

This is closer to the instantaneous speed at X as the time interval involved is very small. This means that:

Instantaneous speed = Average speed of an object – provided the time used is very small.

Measuring instantaneous speed using a computer

The computer uses an internal clock which allows very small time intervals to be measured. This allows the calculation of instantaneous speed if a distance is measured. The computer starts timing when the light beam is cut by the card, and stops when the light beam is restored. The time taken for the card to pass through the beam is recorded in the computer (figure 5.3).

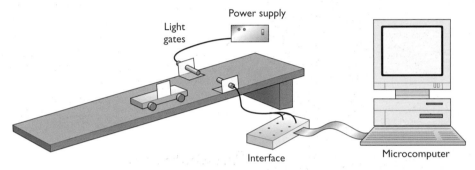

Figure 5.3 Measuring instantaneous speed.

The computer, already pre-programmed with the length of the card, calculates the speed of the vehicle using:

$$\text{Speed} = \frac{\text{length of card}}{\text{time on computer}}$$

Typical results might be:

Length of card = 5 cm = 0.05 m

Time measured by computer = 0.025 s

$$\text{Speed} = \frac{0.05}{0.025}$$

$$= 2.0 \text{ m/s}$$

Fascinating Physics

Measuring car speeds

There are several ways that the speed of moving cars can be measured. The early days of police enforcement of speed limits used a similar method to the one in the lab. A distance was measured and cars were timed as they went between the two points. There was a data table which allowed the police to calculate the speed.

With new technology there are now several ways.

VASCAR

This is the speed system used in police cars. There are two switches, one to measure the time to travel between two markers. The second switch measures the distance between the two markers as covered by the moving

Figure 5.4 (a) Speed camera

Figure 5.4 (b) Doppler mode speed gun.

police car. It is essentially a stopwatch method and calculates average speed but there is no necessity for the police car to travel at the same speed or even in the same direction as the moving car.

Speed guns

These are pointed at a vehicle and can operate in two different ways:
◆ Pulse mode. This uses radar with a wavelength between 0.3 and 1 m. A set of pulses is sent out at a fixed time interval. The machine measures the time between the transmitted and reflected wave. The distance is measured over a number of pulses and the speed can be calculated and displayed.
◆ Doppler mode. Radio waves are sent out and as they reflect back there is a change in frequency. The change is frequency increases as the speed increases.(figure 5. 4(b)).

Speed cameras

These are becoming very common since they do not require any police officers to operate them. A sensor in the camera detects a speeding car and after the car passes by the camera there are two pictures taken half a second apart. There are lines two metres apart on the road. Since the distance between a number of marks and the time taken are known, the speed can be calculated (Figure 5.4(a)).

The cameras are often called Gatso traffic cameras since they were developed by Maurice Gatsonides. He was a former Dutch racing driver.

Acceleration

Table 5.1 gives some information on two vehicles.

Vehicle	Time to go from 0 to 100 km/h in s	Top speed in km/h
A	5.97	270
B	8.0	800

Table 5.1

You could conclude that vehicle A is the faster one since it reaches the speed of 100 km/h in a shorter time. Vehicle B could be the faster since its top speed is greater. Vehicle A is a Porsche motor car and B is a Boeing 737 aircraft. A better way to compare these vehicles is to calculate the acceleration.

Acceleration (a) is the change in speed of an object in one second:

$$\text{Acceleration} = \frac{\text{change in speed}}{\text{time for change}}$$

$$= \frac{\text{final speed} - \text{initial speed}}{\text{time for change}}$$

$$a = \frac{v - u}{t} \quad \text{where } v = \text{final speed;}$$
$$u = \text{initial speed}$$
$$t = \text{time for change in speed to occur}$$

Acceleration is usually calculated in metres per second per second that is m/s^2.

A car has an acceleration of 3 m /s². This means that the speed of the car increases by 3 m/s in every second. If the car starts from rest then:

after 1 s, speed = 3 m/s
after 2 s, speed = 6 m/s
after 3 s, speed = 9 m/s
after 4 s, speed = 12 m/s

Example
A trolley starts from rest and reaches a speed of 8 m/s in 4 seconds. What is the acceleration?

Solution

$$u = 0 \quad v = 8\,\text{m/s} \quad a = ? \quad t = 4\,\text{s}$$

$$\text{Acceleration} = \frac{\text{change in speed}}{\text{time taken}} = \frac{v - u}{t} = \frac{8 - 0}{4} = \frac{0}{4} = 2\,\text{m/s}^2$$

When an object is slowing down it will have a negative value for its acceleration. This is called a **deceleration**.

Example
A pupil is riding a bicycle. She slows down from 8 m/s to 2 m/s in 2 seconds. Calculate the deceleration.

Solution

$$u = 8\,\text{m/s} \quad v = 2\,\text{m/s} \quad a = ? \quad t = 2\,\text{s}$$

$$\text{Acceleration} = \frac{\text{change in speed}}{\text{time taken}} = \frac{v - u}{t} = \frac{2 - 8}{2} = -\frac{6}{2} = -3\,\text{m/s}^2$$

Deceleration = 3 m/s²

Speed – time graphs

From the measurements of speed and time we can draw graphs which allow us to see the motion of an object more clearly. It also allows us to make calculations from the graph to give additional information.

Three types of motion are shown in figure 5.5.

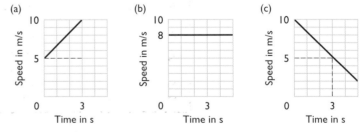

Figure 5.5 Speed-time graphs.

(a) Speeding up. This is a straight line at an angle to the horizontal
(b) Steady speed. This is a straight line parallel to the time axis
(c) Slowing down. This is a straight line heading down towards the time axis.

Important information can be found from the graphs.
◆ The area under any part of a speed–time graph is the distance travelled by the object

◆ To calculate the average speed of an object when more than one type of motion is involved, draw a speed–time graph and find the area under it, i.e. distance travelled, then:

$$\text{Average speed} = \frac{\text{total distance travelled}}{\text{total time taken}}$$

◆ For an object moving only with constant acceleration or constant deceleration:

$$\text{Average speed} = \frac{\text{initial speed} + \text{final speed}}{2}$$

Example

The graph in figure 5.6 shows how the speed of a car varies with time.
(a) Calculate the acceleration.
(b) What is the distance travelled?

Solution

(a)   The acceleration is given as $a = \dfrac{v - u}{t}$

$$\text{where } u = 8\,\text{m/s}$$
$$\text{and } v = 24\,\text{m/s}$$
$$\text{and } t = 4\,\text{s}$$
$$a = \frac{24 - 8}{4} = \frac{16}{4} = 4\,\text{m/s}^2$$

(b)   Distance = Area under (speed–time) ($v - t$) graph

$$= \text{Area 1} + \text{Area 2}$$
$$= 8 \times 4 + \frac{1}{2} \times 16 \times 4$$
$$= 32 + 32$$
$$= 64\,\text{m}$$

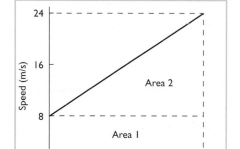

Figure 5.6

Section 5.1 Summary

◆ Average speed $= \dfrac{\text{distance}}{\text{time}}$
◆ Instantaneous speed is calculated the same way as average speed but the time interval is very short.
◆ In both cases the unit is m/s.
◆ Acceleration $= \dfrac{\text{change in speed}}{\text{time taken}} = \dfrac{v - u}{t}$
◆ Unit of acceleration is m/s^2.
◆ Area under a speed–time graph is the distance travelled.

1 **(a)** Calculate the average speed of a car which travels 2400 m in 120 s.

   (b) A car travels 45 m to come to a halt. The average speed of the car is 7.5 m/s. Calculate the time for the car to come to a halt.

2 A bicycle is travelling along a cycle track. Describe how you would measure the instantaneous speed of the cycle.

3 A car magazine states that 'This car accelerates at 2.5 m/s² for the first 10 s.' Explain what is meant by the term acceleration.

4 A car changes its speed from 4 m/s to 14 m/s in 5 s. Calculate the acceleration of the car.

5 The graph below shows how the speed of a vehicle changes with time.

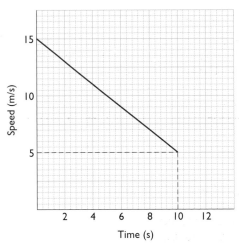

Figure 5.1Q5

(a) Calculate the deceleration of this vehicle.
(b) Calculate the distance travelled by the vehicle.

6 **(a)** A boat starts from rest and reaches a speed of 12 m/s in 18 s. Calculate the acceleration.
   (b) A train accelerates at 6 m/s² for 8 s . It started from rest. Calculate its final speed.
   (c) A car accelerates at 1.5 m/s² from 2 m/s to 14 m/s. Calculate the time for this to happen.

7 For the graphs below, calculate the **(a)** acceleration; **(b)** distance travelled.

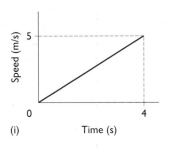 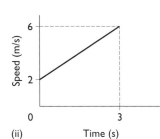

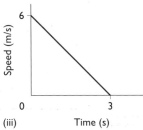 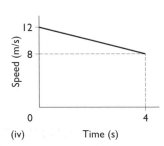

Figure 5.1Q7

At the end of this section you should be able to :

1 Describe the effects of forces in terms of their ability to change the shape, speed and direction of travel of an object.
2 Describe the use of a newton balance to measure force.
3 State that weight is a force and is the Earth's pull on the object.
4 Use the approximate value of 10 N/kg to calculate weight.
5 State that the force of friction can oppose the motion of an object.
6 Describe and explain situations in which attempts are made to increase or decrease the force of friction.
7 State that equal forces acting in opposite directions on an object are called balanced forces and are equivalent to no force at all.
8 State that when balanced forces or no forces act on an object its speed remains the same.
9 Explain, in terms of the forces required, why seat belts are used in cars.
10 Describe the qualitative effects of change of mass or of force on the acceleration of an object.
11 Carry out calculations involving the relationship between a, F and m.
12 Distinguish between mass and weight.
13 State that the weight per unit mass is called the gravitational field strength.
14 Explain the movement of objects in terms of Newton's first law.
15 Carry out calculations using the relationship between a, F and m and involving more than one force but in one dimension only.

The effects of forces

When an object is pushed or pulled, a force is exerted on it.
◆ A force can make a stationary object move.
◆ A force can change the speed of a moving object.
◆ A force can change the direction of a moving object.
◆ A force can change the shape of (deform) an object.
These effects depend on the size of the force applied to the object.

Measuring force

Springs can be used to measure force:
1 A spring stretches evenly – each time a mass is added to the carrier the spring stretches by the same amount.
2 A spring returns to its original length when the force is removed. If the force is too great the spring will not return to its original length when the force is removed.
We measure forces using a Newton balance in units of **newtons (N)**. (figure 5.8).

Frictional forces

Objects such as cars can slow down due to forces acting on them. These forces can be due to the road surface and the tyres or due to the brakes. The force that tries to oppose motion is called the **force of friction**. A frictional force always acts when particles are sliding across one another and can oppose any motion. This force will depend on the nature of the surface. It

Figure 5.7 The force of the kick deforms the ball.

Figure 5.8 A Newton balance.

will try to slow down an object like a car whether it is going in either direction along a road.

Car brakes

Car brakes operate by slowing down the car. When the brake pedal is pressed it causes the brake shoes to be pushed against the drum.

Some cars are fitted with an anti-lock braking system, called ABS, which has sensors located at each wheel to detect when it is about to lock during braking. If the brakes lock it will be very difficult to steer. It prevents this, and so the risk of skidding, by rapidly releasing and reapplying the brake (figure 5.9).

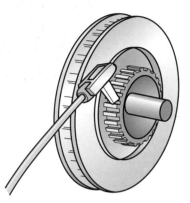

When the sensor detects that there is wheel lock, it reduces brake pressure and then reapplies the brakes.

Figure 5.9 How an ABS system works.

Reducing friction

Lubrication

The frictional force between two surfaces moving against each other can be reduced by lubricating the surfaces. This generally means that oil can be placed in between two metal surfaces. This happens in car engines and reduces wear on the engine, since the metal parts are not actually meeting each other but have a thin layer of oil.

Streamlining

The reduction in the flow of a fluid over a surface is called **streamlining**. This can occur in both vehicles and people.

Modern cars are designed to offer as little resistance (drag) to the air as possible. This is the friction of the air on the car. To reduce this friction the designers try to streamline the vehicle in a variety of ways. This streamlining is measured by a number called the **drag coefficient**, C_d. The larger the C_d number, the greater the resistance to air flow. The calculation of C_d is difficult and involves the vehicle being placed in wind tunnels and smoke flowing over it as shown in figure 5.10. Most C_d values range from 0.3 to 0.4.

The drag on the car can be reduced in a number of ways:
◆ reducing the front area of the car;
◆ not carrying a roof rack;
◆ having door mirrors instead of wing mirrors;
◆ having a smooth round body shape;
◆ having aerials made as part of the car windows.

Smooth designs

When you try to move fast through water as a swimmer it helps to have as smooth an outline as possible. The modern design of swimwear produces a smooth shape so that the disturbance of the water is reduced to a minimum (figure 5.11).

Figure 5.10 Measuring the drag coefficient of a car.

Figure 5.11 Design of swimwear helps to make the swimmer streamlined.

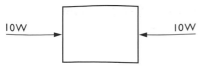

Figure 5.12

Balanced and unbalanced forces

When two or more forces act on an object, the combined effect depends on their size and direction.

Balanced forces acting on an object are equal in size but act in opposite directions. They cancel each other out and so are the equivalent of zero force acting on the object (figure 5.12) An unbalanced force acting on a body cause it to speed up or slow down.

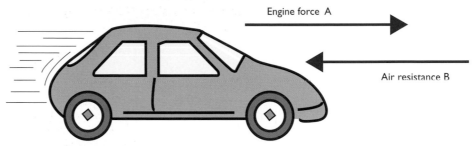

Figure 5.13 Balanced forces.

In figure 5.12, when *A* and *B* are both the same, the same force is applied to each side of the vehicle, i.e. the forces are balanced. If the vehicle is at rest it will stay at rest. If moving, it will continue to move at a constant speed in a straight line.

When the engine force *A* is greater than the air resistance force *B*, the car will accelerate to the right.

Newton's First Law

Figure 5.14 Sir Isaac Newton.

Sir Isaac Newton is regarded as one of the greatest scientists. He was born in 1642 in Grantham and studied at home and school and later went up to Cambridge. At the age of 24 he was appointed Lucasian Professor of Mathematics. He worked on mathematics, light and astronomy and proposed three laws of motion (figure 5.13)

This is his First Law:

> A body will remain at rest or move at constant speed in a straight line unless acted on by an unbalanced force.

Using Newton's First Law, we can explain the following motions:
1 A speed boat travelling at a steady speed through the water (figure 5.15) The resistive force offered by the water will balance out the force exerted by the engine.
2 A car moving at a constant speed along a level road and no matter how hard the driver tries to accelerate the car will not increase its speed (see figure 5.13). Again the resistive forces produced by the air resistance force will balance the force produced by the engine.
3 Blood rushes from your head to your feet when you quickly come to a halt as a lift descends. This happens because the blood keeps on moving down at a constant speed in a straight line.
4 When you wish to dislodge sauce from the bottom of a bottle you turn it upside down and throw it downwards very fast and then bring it to a halt. This happens because the sauce keeps moving down at constant speed in a straight line.
5 Headrests prevent injuries called whiplash during collisions from the rear since they prevent the rapid movement of the head.

Figure 5.15

Force, mass and acceleration

If a mass m is accelerated by a force (F), if we double the force we double the acceleration. If you keep the force constant and double the mass, then you will halve the acceleration. This can be written as Newton's Second Law, namely:

> When an object is acted on by a constant unbalanced force, the body moves with constant acceleration in the direction of the unbalanced force.

$$\text{Unbalanced force} = \text{mass} \times \text{acceleration}$$

In symbols, $F = ma$, where F is in newtons (N); m is in kilograms (kg); and a is in metres per second squared (m/s^2).

Example
A constant unbalanced force of 8 N acts on a mass of 10 kg. Calculate the acceleration of the 10 kg mass.

Solution
$$F = 8\,\text{N} \quad m = 10\,\text{kg} \quad a = ?$$
$$F = ma$$
$$a = \frac{F}{m} = \frac{8}{10} = 0.8\,\text{m/s}^2$$

Example
A car on a fun ride is being pulled by a constant unbalanced force of 600 N. The car accelerates at 15 m/s^2. What is the mass of the car?

Solution
$$F_{un} = 600\,\text{N}$$
$$a = 15\,\text{m/s}^2$$
$$600 = m \times 15$$
$$m = \frac{600}{15}$$
$$= 40\,\text{kg}$$

Example
A toy car is being pulled by a constant force of 20 N. The car has a mass of 2 kg. There is a constant frictional force of 4 N acting on the car. Calculate the acceleration of the car (figure 5.16).

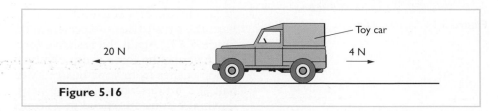

20 N   Toy car   4 N

Figure 5.16

Solution
The unbalanced force acting on the car is:

$$F_{un} = 20 - 4 = 16\,\text{N}$$
$$a = \frac{F_{un}}{m} = \frac{16}{2} = 8\,\text{m/s}^2$$

Mass and weight

We often talk about the weight of a person being 70 kg when what a physicist would mean is mass. The distinction may not be obvious. Firstly let's talk about mass.

Mass

Mass is the quantity of matter forming an object. Mass depends on the number and type of atoms making it up, so the mass of an object remains constant. If you go to different parts of the Earth or even to the Moon you have the same mass, since hopefully you do not lose an arm or a leg on the journey.

Force of gravity

Force of gravity is the downward pull of the Earth on an object.

$$\text{Force of gravity} = \text{pull of Earth on an object}$$
$$= \text{gravitational force on an object}$$
$$= \text{weight of object}$$

Weight

The weight of an object depends on: (a) mass, and (b) where it is. Being a force, weight is measured in newtons (N).

$$\frac{\text{weight of object}}{\text{mass of object}} = \frac{W}{m} = \text{constant}$$

This constant is called the gravitational field strength (g) and is measured in N/kg.

Gravitational field strength varies depending on where you are. For instance, on Earth $g = 10$ N/kg but on the Moon $g = 1.6$ N/kg and on Jupiter $g = 26.4$ N/kg. This information is shown on the data page.

Thus:

$$\text{Weight} = \text{mass} \times \text{gravitational field strength}$$
$$W = mg$$

Example
What is the weight of 4 kg mass?
Solution

$$W = mg$$
$$= 4 \times 10$$
$$= 40\,\text{N}$$

Acceleration due to gravity

The force of gravity causes all objects to accelerate as they fall back to Earth, since the Earth is so massive and exerts a very strong force on these small objects. This gravitational force is the one force that we cannot switch off.

For an object falling freely:

$$\text{Unbalanced force } F = \text{weight} = \text{mass} \times \text{acceleration}$$

$$W = ma$$

$$\text{And } a = \frac{W}{m} = \text{acceleration due to gravity}$$

$$\text{Gravitational field strength} = \frac{\text{force of gravity on a body}}{\text{mass of body}}$$

But gravitational field strength $= \dfrac{W}{m} =$ acceleration due to gravity.

This means that the units of g are N/kg or m/s^2.

Forces and supported bodies

We know from Newton's First Law that if an object either remains stationary or moves upwards or downwards at constant speed, then the upward force is equal in size to the weight of the object but acts in the opposite direction.

Example

A stationary mass m is hanging from a string. The weight of the object is mg (N) and acts downwards. This is balanced by a force of the same size acting upwards due to the tension in the string (figure 5.17).

Example

A book rests on a shelf which gives an upward force, called the reaction force, equal in size to the weight, therefore the forces are balanced (figure 5. 18).

We can also see an example of balanced forces in action in different situations.

Aircraft

The engines provide a forward force or **thrust** which accelerates the aircraft forward. However, as it moves faster the air resistance, or drag, increases until the forces (horizontally) are balanced. The aircraft would move at constant speed (Newton's First Law). When in level flight at constant speed the increased *lift* will balance the weight giving balanced forces (figure 5. 19).

Powerboat

Air offers less resistance to motion than water. A boat uses most of its power to push its hull through the water. This is because the force of friction due to the water is large.

Inertia

All objects are unwilling (or reluctant) to change their motion. This means that a steady force is required to change the motion of an object.

This reluctance of an object to change its motion is called its **inertial mass** or mass of inertia. It depends on mass. The larger the mass, the larger the inertia and the more unwilling the body is to change its motion. It is easier to stop a lighter object than a more massive one when both are travelling at the same speed.

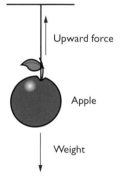

Upward force

Apple

Weight

Figure 5.17

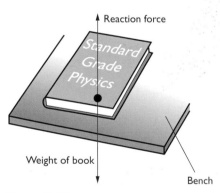

Reaction force

Standard Grade Physics

Weight of book

Bench

Figure 5.18

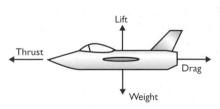

Lift

Thrust

Drag

Weight

Figure 5.19

Seat belts – or why Newton's laws may save your life

When a car brakes suddenly, any unrestrained object will continue to move at the car's original speed (Newton's First Law). It will probably collide with some part of the interior causing damage or injury.

A seat belt applies a force in the opposite direction of motion causing rapid deceleration of the wearer. The webbing straps are designed to have a certain amount of 'give' so that the sudden force applied to the person does not cause injury (figure 5.20). Adult seat belts are not suitable for young children since the young children tend to slip down in the seat. This effect is called 'submarining' and can be prevented by using a special design for children.

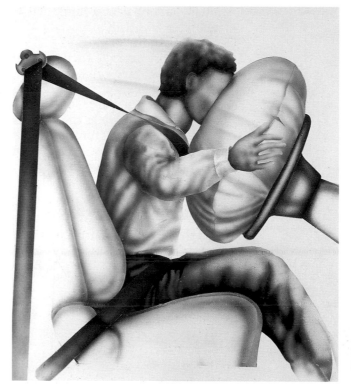

Figure 5.20 Motion of a car seat belt.

Fascinating Physics

Increasing time to collision

If you travel at 60 km per hour ,which is about 17 m/s, then every 5 km per hour increase doubles the risk of a fatal crash. It is useful to consider the forces acting on a human. If you are travelling at 2 m/s and hit a solid object then the time taken to come to a halt is about 0.01 s. For a mass of 70 kg, the force acting on you is 14 000 N. If the time is increased to 0.2 s then the force is reduced to 700 N. Even though this is still a large force you are likely to survive rather than suffer severe injuries. In order to increase the time cars can use an air bag, which is now compulsory in all new cars. The large area of the bag also prevents chest injuries due to the steering wheel (figure 5.21).

The second feature in cars that will increase the time is the car's 'crumple zone'. The main compartment in a car is a safety 'cage', that is a rigid structure to protect the occupants. The front and rear of the car will collapse relatively easily and increase the time for the collision which decreases the force.

Figure 5.21

Frictional forces in a fluid

A fluid is a liquid or gas. In a fluid, the frictional force on an object travelling through it increases as the speed of the object increases. This can be shown by dropping metal spheres into tubes of a thick oily liquid and water.

In both cases the spheres accelerate and then move with constant speed (**terminal velocity**). When the sphere is accelerating there is an unbalanced force acting on it. When the terminal velocity is reached the forces are balanced. It is reached sooner in the thick liquid than in water. This shows that the frictional force in a liquid depends on the type of liquid.

The motion of an object falling through any fluid can be divided into three parts:

1. Initially an unbalanced force acts on the object due to its weight and the object falls with a constant acceleration of $10\,\text{m/s}^2$, which is the acceleration due to gravity.
2. After a short time the frictional force begins to act and alters the motion. This force will be increasing as the speed of the object increases. There is a smaller unbalanced force ($F_{un} = W - F_r$) and the acceleration is therefore less than $10\,\text{m/s}^2$. This acceleration will continue to decrease as the frictional force increases.
3. Finally the frictional force balances the weight. We now have balanced forces. The object now falls at a constant speed in a straight line. The object has reached its greatest speed called its terminal speed or terminal velocity. (figure 5.22).

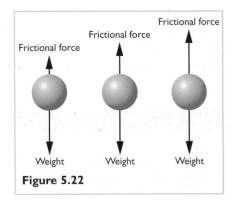

Figure 5.22

Free falling parachutist

The graph of speed against time for a parachutist falling in free fall out of an aircraft and then opening the rip cord some time later, is shown in figure 5.23. Each part of the graph is explained below:

Figure 5.23 (a) The sky diver has reached her terminal velocity.

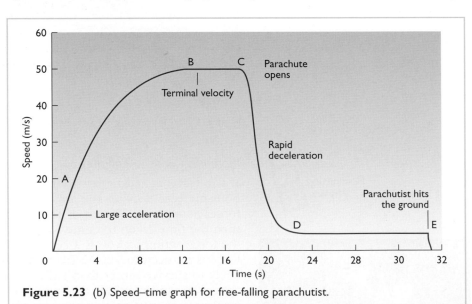

Figure 5.23 (b) Speed–time graph for free-falling parachutist.

OA = constant acceleration due to gravity.
AB = decreasing acceleration as frictional force acts.
    Unbalanced force = weight – frictional force.
BC = constant speed as frictional force upwards = weight downwards.
CD = non-uniform deceleration due to parachute opening and increasing frictional force.
DE = constant speed as frictional force upwards = weight downwards.
EF = parachutist hits the ground.

Section 5.2 Summary

◆ A force can change the shape, speed and direction of travel of an object.
◆ A Newton balance is used to measure force.
◆ Weight is a force and is the Earth's pull on the object.
◆ Mass is the amount of matter in an object.
◆ The approximate value of g the Earth's gravitational force on a unit mass is 10 N/kg and is the weight per unit mass.
◆ The force of friction can oppose the motion of an object.
◆ Equal forces acting in opposite directions on an object are called balanced forces and are equivalent to no force at all.
◆ When balanced forces or no forces act on an object its speed remains the same. This is Newton's First Law.
◆ Seat belts are used in cars to exert an opposite force.
◆ Newton's Second Law can be written as $F = ma$.

End of Section Questions

1   A pupil has a mass of 50 kg and sits on a bicycle of mass 20 kg. If the accelerating force is 140 N, calculate the acceleration of the pupil and cycle.

2   A mass of 5 kg accelerates at 6 m/s². Calculate the unbalanced force required.

3   A car of mass 800 kg travelling at 18 m/s sees an obstacle ahead.

   The driver brakes to a halt in 9 s.

   (a) Calculate the acceleration
   (b) Calculate the unbalanced force acting on the car.

4   A rocket has an acceleration of 40 m/s². The mass of the rocket is 10 000 kg. If the force exerted by the engines acting on the rocket is 405 000 N, calculate the frictional force acting on the rocket.

5   A student has a mass of 50 kg. What will her weight be (a) on Earth? (b) on the moon?

6   Explain, using the correct law of physics, how seat belts produce less injury during an accident.

7   During a car check the tyres of a car are noticed to have less than the minimum tread required by law. What is the necessary feature of the tread in producing grip?

At the end of this section you should be able to:

1 Describe the main energy transformations as a vehicle accelerates, moves at constant speed, brakes and goes up or down a slope.
2 State that work done is a measure of the energy transferred.
3 Carry out calculations involving the relationship between work done, force and distance.
4 Carry out calculations involving the relationship between power, work and time.
5 State that the change in gravitational potential energy is the work done against/by gravity.
6 State that the greater the mass and/or the speed of a moving object, the greater is its kinetic energy.
7 Carry out calculations involving the relationship between kinetic energy, mass and speed.
8 Carry out calculations involving energy, work, power and the principle of conservation of energy.

Work and energy

Energy is a very useful quantity. Simply, energy allows objects to move and lets you do things – it is called **work**.

Energy has many different forms. When work is done, energy is transferred to an object or changed into another form.

> **Energy transferred = work done**
> **= applied force × distance moved by the force**

Thus

$$\text{Work done (Wd)} = F \times d$$
where d distance in metres, and F = applied force in newtons.

The unit of energy and work done is the Joule (J), thus:

$$1 \text{ joule} = 1 \text{ newton metre}$$
$$1\,\text{J} = 1\,\text{N m}$$

Example
A boy finds he has to exert a force of 50 N to lift a box 2 m onto a shelf. Calculate the work done by the boy.
Solution

$$\text{Work done} = F \times d \quad \text{where } F = 50\,\text{N and } d = 2\,\text{m}$$
$$\text{Work done} = 50 \times 2$$
$$= 100\,\text{J}$$

Conservation of energy

Energy cannot be created or destroyed but can be changed from one form to another, when work is done. For example, the kinetic energy of movement in a car is changed mainly into heat when the brakes are applied. If a system gains energy then another system loses the same amount of energy. This is called the conservation of energy.

Gravitational potential energy (E_p)

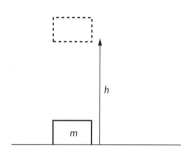

Figure 5.24

Water in a mountain loch has stored energy (gravitational potential energy) which can be transferred into electrical energy in a hydroelectric scheme. It is available as the water is above the generating station and can be transferred by allowing it to fall.

We can derive an equation to allow us to calculate the potential energy for a mass being lifted. A mass, m, is lifted at constant speed through a vertical height of h metres. (figure 5.24).

The work done in lifting it is calculated as:

$$\text{Work done} = F \times d$$

In this case the force applied must balance the weight of the box.

$$\text{Applied force upwards} = \text{weight downwards}$$
$$F = mg$$
$$\text{Work done} = mg \times h$$
$$= mgh$$
$$= \text{gain in gravitational potential energy } (E_p)$$

$$E_p = mgh \quad \text{where} \quad \begin{aligned} E_p &= \text{change in gravitational } E_p \text{ (J)} \\ m &= \text{mass (kg)} \\ g &= \text{gravitational field strength (N/kg)} \\ h &= \text{vertical height } (m) \end{aligned}$$

Kinetic energy (E_k)

This is the energy possessed by moving objects. We can also derive an equation for this type of energy.

Consider an object of mass, m, starting from rest. This object is accelerated by a force, F. The final speed, v, is reached in a time, t (figure 5.25). The effects of friction can be neglected.

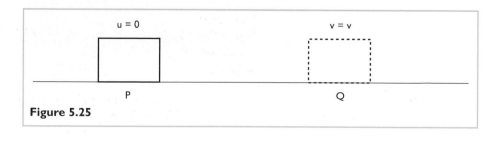

Figure 5.25

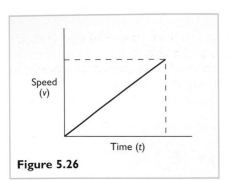

Figure 5.26

Work done on object $= F \times d$

$\qquad\qquad\qquad = ma \times$ Area under speed–time graph (figure 5.26)

and $a = \dfrac{v - u}{t} = \dfrac{v - 0}{t} = \dfrac{v}{t}$

Area under the graph $= \dfrac{1}{2} \times t \times v = \dfrac{1}{2}vt$

Work done on object $= \dfrac{mv}{t} \times \dfrac{1}{2}vt$

$\qquad\qquad\qquad = \dfrac{1}{2}mv^2$

Work done on object = Gain in E_k

$\qquad\qquad E_k = \dfrac{1}{2}mv^2$

Note that E_k depends on v^2. This means that:
◆ If the speed doubles then the kinetic energy increases by four.
◆ If the speed trebles then the kinetic energy increases by nine.

Power

When we say that an appliance is more powerful than another appliance, we mean that it uses up energy at a faster rate.

Power = energy transferred in 1 s = work done in 1 s

$$\text{Power} = \frac{\text{energy transferred}}{\text{time taken}} = \frac{\text{work done}}{\text{time}}$$

Power is measured in watts (W). A power of one watt means that one joule of energy is transferred in one second.

Example
A 5 kg box is raised through 20 m in 4 s.
(a) Calculate the gain in gravitational E_p of the box.
(b) Calculate the power used to lift the box.
(c) Explain why more power would be needed than the value calculated in (b).

Solution

(a) $\qquad\qquad\qquad\qquad E_p = mgh$

$\qquad\qquad\qquad\qquad\quad = 5 \times 10 \times 20$

$\qquad\qquad\qquad\qquad\quad = 1000\,\text{J}$

(b) $\qquad\qquad\qquad\qquad P = \dfrac{E}{t}$

$\qquad\qquad\qquad\qquad\quad = \dfrac{1000}{4}$

$\qquad\qquad\qquad\qquad\quad = 250\,\text{W}$

(c) Some of the energy supplied is changed into heat and not into lifting the box.

Example

A 50 kg girl on a 15 kg bicycle is moving at a uniform speed of 5 m/s when she applies the brakes and comes to rest in 2 seconds.

(a) What is the kinetic energy of the girl and her bicycle before she brakes?

(b) What becomes of this kinetic energy during the braking?

(c) Calculate the power of the brakes.

Solution

(a)
$$E_k = \frac{1}{2}mv^2$$
$$= \frac{1}{2} \times 65 \times (5)^2$$
$$= 813 \, \text{J}$$

(b) **It changes into heat.**

(c) Change in E_k = 813 J = energy transferred

$$P = \frac{E}{t}$$
$$= \frac{813}{2}$$
$$= 407 \, \text{W}$$

Estimating your own power

Figure 5.27

In going upstairs, you will gain in gravitational potential energy. An applied force equal to your weight (force of gravity) acts vertically upwards and so the distance moved by the applied force is the vertical height of the stairs.

You can measure your weight on bathroom scales (figure 5. 27). To calculate your power you also need the time for you to go up the stairs.

Run up the stairs and time how long it takes you to run. Measure the vertical height of the stairs.

Typical results might be:

$$\text{Height of stairs} = 3 \, \text{m}$$
$$\text{Time taken to go up stairs} = 3.5 \, \text{s}$$
$$\text{Weight} = 700 \, \text{N}$$

Work done going up stairs = energy transferred
= gain in gravitational potential energy
= *mgh*
= 700 × 3
= 2100 J

$$\text{Power} = \frac{\text{energy transferred}}{\text{time taken}} = \frac{mgh}{t} = \frac{2100}{3.5}$$
$$= 600 \, \text{W}$$

The power of a horse is about 750 W and an Olympic runner can produce up to 3000 W. We cannot sustain this power for any length of time. Most of the power loss occurs as heat loss as we sweat rapidly during such a large output of power.

Safety and design in moving vehicles

Every year there are many people killed or injured as a result of motor accidents. Some of these accidents can be prevented by better driving particularly in bad weather conditions. However, accidents will occur, and in attempting to avoid an accident, there are only two things a driver can do – change the car's speed and/or its direction.

In an accident, the design of the car and of the roadside environment can greatly determine the amount of damage and injury which might occur. If there were no protective devices or designs in a car then you would continue to move at the same speed as the car if the car is struck or suddenly comes to a halt. This would result in your going through the windscreen or being injured by the sides of the door being crushed against you.

The roadside environment

Twenty-five per cent of road accidents are with roadside objects so safety rails and posts are now designed to deform on impact and absorb most of the energy of a vehicle, reducing the effects of an accident. Roadside barriers are made of metal sheets which absorb some of a vehicle's energy when struck. More energy is absorbed by the posts which 'give' since they are not buried deep in the ground.

Many road accidents are caused by speed. Doubling the speed of the car gives it four times the kinetic energy. This kinetic energy has to be used up in such an accident, consequently lower speeds can dramatically reduce the damage to the car involved in a road accident. In the USA the maximum speed limit on motorway type roads was 55 miles per hour (mph). This reduction in speed was introduced to reduce fuel consumption but should also help reduce damage and injury in accidents.

Thinking and braking

Reaction time is the time between seeing and reacting to an event. Reaction times vary from person to person. An average driver has a reaction time of about 0.8 s when this is measured under normal conditions of driving. Reaction time is likely to be much longer if drugs or alcohol have been taken. Even a small amount of alcohol can greatly increase reaction time.

Thinking distance is the distance travelled during this reaction time.

$$\text{Thinking distance} = \text{speed} \times \text{reaction time}$$

Braking distance is the distance travelled from the time the brakes were applied to the vehicle becoming stationary.

Stopping distance consists of two parts, the thinking distance and the braking distance which are added together. The stopping distance for a car travelling at 30 mph in good road conditions is 23 m and at 60 mph it is 73 m. These shortest stopping distances are shown in table 5.2.

Speed (mph)	Thinking distance (m)	Braking distance (m)	Overall stopping distance (m)
20	6	6	12
30	9	14	23
40	12	24	36
50	15	38	53
60	18	54	72
70	21	75	96

Table 5.2 Shortest stopping distances at various speeds.

Since thinking distance is speed × time and the reaction time remains constant, the thinking distance increases by the same amount when the speed increases.

Energy conservation – gravitational E_p to E_k

During any energy transformation the total amount of energy is always conserved, that is it stays the same, but may be changed into less useful forms. In a motor car the chemical energy of the fuel is changed into kinetic energy of movement and heat energy in the exhaust fumes. This heat energy is not useful and is contributing to the concerns about global warming.

Energy transfer or energy changes

When work is done on a system, energy is transferred to or from that system.
(a) If the speed of an object changes, then, from the conservation of energy:

$$\text{Work done} = \text{change in } E_k$$
$$\text{Force applied} \times \text{distance moved by force} = E_k \text{ (large)} - E_k \text{ (small)}$$

(b) If the height of an object changes, then, from the conservation of energy:

$$\text{Work done} = \text{change in gravitational } E_p$$
$$\text{Force applied} \times \text{distance moved by force} = mgh$$

Examples of energy transfer

The pendulum
Figure 5.28 shows a pendulum bob moving from B to A. As it moves, work is done and energy transferred to the bob.

$$\text{Work done on bob} = \text{energy transferred to bob}$$
$$= \text{gain in gravitational } E_p$$
$$= mgh$$

When the bob swings from A to B it 'loses' gravitational potential energy, E_p, but gains kinetic energy, E_k. At B it has maximum speed and hence maximum E_k.
 If no energy is transferred with the surroundings then:

$$\text{Loss in gravitational } E_p = \text{gain in } E_k$$

$$mgh = \frac{1}{2} mv^2 - 0$$

$$mgh = \frac{1}{2} mv^2$$

If friction is present, the bob still loses the same amount of gravitational E_p. This is now transferred into E_k and work is done against friction. The work done against friction may be in the form of heat and sound energies, which are 'lost' or dissipated from the system:

i.e. Loss in gravitational E_p = gain in E_k + work done against friction

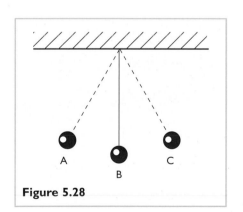

Figure 5.28

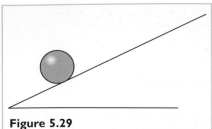

Figure 5.29

Sledge on a slope (figure 5.29)

If a sledge is *projected up the slope*, it gets slower and slower and so is losing E_k, but gaining gravitational E_p.

If no energy is transferred with the surroundings, then:

$$\text{Loss in } E_k = \text{gain in } E_p$$
$$\frac{1}{2}mv^2 - 0 = mgh$$

If friction is present on the incline then work will be done against friction:

$$\textbf{Loss in } E_k = \textbf{gain in gravitational } E_p + \textbf{work done against friction}$$

$$\frac{1}{2}mv^2 - 0 = mgh + F_r \times d$$
$$\frac{1}{2}mv^2 = mgh + F_r \times d$$

The falling body sky diving without the parachute

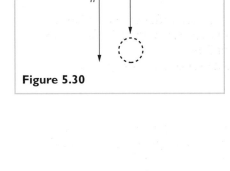

Figure 5.30

As the object falls it 'loses' gravitational E_p but gains E_k. Just before impact with the ground it has 'lost' all its gravitational E_p and has only E_k (figure 5.30).

If no energy is transferred with the surroundings, then:

$$\text{Gain in } E_k = \text{loss in gravitational } E_p$$

$$\frac{1}{2}mv^2 - 0 = mgh$$
$$\frac{1}{2}mv^2 = mgh$$

If the object is *projected vertically upwards*, for example the human cannonball, then it slows down and so loses E_k and in gaining height has gained gravitational E_p.

If no energy is transferred with the surroundings, then:

$$\text{Loss in } E_k = \text{gain in gravitational } E_p$$

$$\frac{1}{2}mv^2 - 0 = mgh$$
$$\frac{1}{2}mv^2 = mgh$$

Example

A pendulum swings as shown in figure 5.28. Points A and C are the extremities of the swing. The mass of the bob is 0.5 kg. The maximum vertical height reached is 0.1 m.

For the bob:
(a) Calculate the change in gravitational potential energy.
(b) Calculate the maximum kinetic energy.
(c) Calculate the top speed.

Solution

(a)
$$E_p = mgh$$
$$= 0.5 \times 10 \times 0.1$$
$$= 0.5 \, \text{J}$$

(b) Since we assume that all the potential energy changes into kinetic energy:

$$E_k = E_p = 0.5\,\text{J}$$

(c)
$$\tfrac{1}{2}mv^2 = 0.5$$

$$\tfrac{1}{2}(0.5)v^2 = 0.5$$

$$v^2 = 2$$

$$v = 1.44\,\text{m/s}$$

Example

When Galileo dropped metal spheres of different mass from the leaning tower of Pisa he found that they hit the ground at the same time. How can his discovery be explained in terms of the conservation of energy?

Solution

If the change in gravitational potential energy equals the change in kinetic energy then:

$$mgh = \tfrac{1}{2}mv^2 - 0$$

Then the masses will cancel on both sides of the equation giving:

$$gh = \tfrac{1}{2}v^2$$

Since the height and speed are connected but are independent of mass then any mass will fall and hit the ground at the same time. This is only true if the effects of air resistance are negligible.

Galileo did this experiment but the initial ideas on gravity were put forward by Sir Isaac Newton.

Example

The tower at the Glasgow Science Centre has a height of 104 m. If an object of mass 5 kg is dropped, to check the laws of physics, estimate the speed of the object at the bottom of the tower (figure 5.31).

Solution

Using
$$gh = \tfrac{1}{2}v^2$$

$$10 \times 104 = \frac{v^2}{2}$$

$$v^2 = 2008$$

$$v = 44.8\,\text{m/s}$$

Figure 5.31 Glasgow Science Centre.

- Work done is a measure of the energy transferred.
- Work done = force × distance and is measured in joules (J).
- Power = $\dfrac{\text{work done}}{\text{time}}$ and is measured in watts (W).
- Gravitational potential energy is the work done against gravity, $E_p = mgh$
- Kinetic energy $E_k = \frac{1}{2}mv^2$

End of Section Questions

1 (a) A sledge is pulled with a force of 15 N for a distance of 200 m. Calculate the work done.
   (b) A box is pushed 12.5 m along a floor and 100 J of energy are used. Calculate the force needed for this work.
   (c) A force of 60 N is needed to push a bicycle and 360 J of energy are used. How far does the bicycle move ?

2 A 25 kg container is pushed 20 m using a force of 35 N. The time taken is 60 s.

   (a) Calculate the work done.
   (b) If this occurs in 60 s calculate the power used.

3 A weightlifter lifts a mass of 100 kg a height of 1.5 m. Calculate the potential energy gained.

4 A mass of 10 kg gains a potential energy of 400 J. How high must the mass have been raised?

5 A water pump can pump 100 kg of water in 1 minute through a height of 2 m. Calculate the power of the pump.

6 During a game a ball of mass 1.5 kg travels at 5 m/s. Calculate the kinetic energy of the ball.

7 A student runs at a constant speed of 4 m/s and has a kinetic energy of 400 J. Calculate the mass of the student.

8 A skater of mass 60 kg comes to a halt in a distance of 5 m against a frictional force of 3 N. How much kinetic energy did the skater use?

9 A stone of mass 0.05 kg is thrown up with a speed of 12 m/s.

   (a) Calculate the kinetic energy of the stone.
   (b) Calculate the height reached by the stone.

1 A student observes some measurements of speed trials on cars. The engineer states that she measures the instantaneous speed rather than the average speed.

(a) Explain how you could measure the instantaneous speed of a car on a road. In your description you should state:
  a. the apparatus you would use;
  b. what measurement you would make;
  c. how you would use those measurements to calculate the instantaneous speed.
(b) The car has a length of 4.2 m and takes 0.4 s to pass the student. Calculate the instantaneous speed of the car.

2 During a trial of a new car, model A reaches 16 m/s from rest in 8 s.

(a) Calculate the acceleration of this car.
(b) Model B car takes 10 s to go from 2 m/s to 16 m/s. By calculating its acceleration, state which car has the greater acceleration.
(c) If car B has a mass of 800 kg, calculate its kinetic energy when it reaches 16 m/s.

3 A student is measuring the frictional force on a motorbike by pulling it with a Newton meter. Another student sits on the bike.

(a) If the bike is pulled at constant speed, how do the forces acting on the bike compare?
(b) If the force on the balance is 80 N and the bike travels a distance of 12 m, calculate the work done on the bike by the student.

4 A boat has a mass of 150 000 kg.

(a) What is the weight of the boat?
(b) The boat glides easily through the water leaving as little disturbance as possible. What is the name of the feature of the design of the boat which achieves this effect?

5 A shopping trolley has a mass of 14 kg when fully loaded. An energetic student pulls it to accelerate at 1.5 m/s^2. The frictional force acting on the trolley is 5 N.

(a) (i) Calculate the unbalanced force acting on the trolley.
  (ii) What is the total force produced by the student?
(b) The trolley reaches a speed of 5 m/s and then travels at this constant speed for a short time. Calculate the kinetic energy at this time.

6 During construction of a building a lift is used to bring equipment up several floors.

(a) The equipment has a mass of 60 kg and is lifted a height of 12 m. Calculate the potential energy of the equipment.
(b) This operation takes place in 3 s. Calculate the power of the lift.
(c) The actual power produced by the lift motor is greater than the answer to (b). Explain why this is the case.

CHAPTER SIX

Energy Matters

At the end of this section you should be able to:

1 State that fossil fuels are at present the main sources of energy.
2 State that the reserves of fossil fuels are finite.
3 Explain one means of conserving energy related to the use of energy in industry, in the home, and in transport.
4 Carry out calculations relating to energy supply and demand.
5 Classify renewable and non-renewable sources of energy.

6 Explain the advantages and disadvantages associated with at least three renewable energy sources.

Industrialised countries, like Britain, use large quantities of energy for heating, transport and industry. This energy comes largely from the **fossil fuels**: coal, oil, and gas. They are called fossil fuels because they are the remains of plants and animals which lived many millions of years ago. Once they have been used up, they cannot be replaced, i.e. they are finite. Coal is mainly used to produce electrical energy at power stations. Britain has sufficient coal reserves to last about 300 years.

Oil is used mainly for transport. World supplies of oil are expected to last about 60 years. Gas is largely used for space heating (heating homes) and for generating electricity. Britain's North Sea oil and gas supplies will largely be used up by about the year 2030. Burning fossil fuels produces large quantities of poisonous gases, such as sulphur dioxide, and carbon dioxide. Sulphur dioxide is one of the of gases responsible for acid rain. Carbon dioxide, a greenhouse gas, is one of the major causes of global warming.

Because fossil fuels will run out, cause pollution and are expensive, it is important to use them efficiently. In addition, alternative sources of energy are slowly being developed to take the place of fossil fuels.

Energy conservation

Government agencies and the gas and electricity companies all produce literature which explains how we can all save energy (and therefore money) in our homes. Some examples are fitting a hot-water tank jacket, reducing draughts through windows and doors, loft insulation and the replacement of filament lamps with energy-efficient lamps.

There are a number of ways that energy can be saved in our everyday lives. In transport: by sharing a car, using public transport instead of a car, walking or cycling. In industry: by using energy-efficient lighting, heating only the parts of a factory that are being used, and installing energy-efficient machinery.

Power stations

In a conventional power station, coal, gas or oil is burnt in a furnace and the heat produced changes water into steam in a boiler. The steam produced in the boiler is at high pressure. It drives the **turbine** at high speed and this drives the **generator** which produces electrical energy. Steam leaving the turbine enters the **condenser** which turns it back into water. The water is then pumped back to the boiler under pressure.

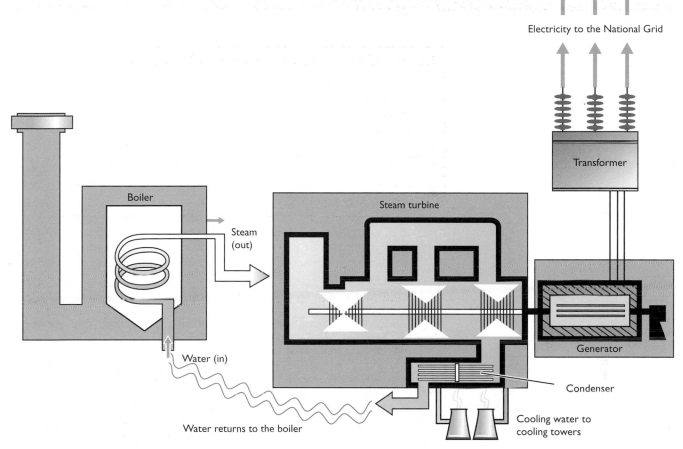

Figure 6.1 A conventional power station.

Cooling towers are required to cool the large quantities of water needed by the condenser (figure 6.1). Only about 35 per cent of the fuel's energy (the energy input to the power station) is converted into electrical energy, the remaining 65 per cent is 'lost' as waste heat. A large proportion of this waste heat is transferred to the atmosphere by the cooling towers.

A **combined heat and power** (CHP) station not only generates electricity but also supplies hot water to heat buildings in district heating schemes. The hot water is piped to local houses and other buildings, passed through radiators and returned, cooler, to the power station for re-heating. About 25 per cent of the energy input to the power station is converted into electrical energy and 50 per cent is usefully used to heat homes and offices, leaving about 25 per cent as waste heat (figure 6.2).

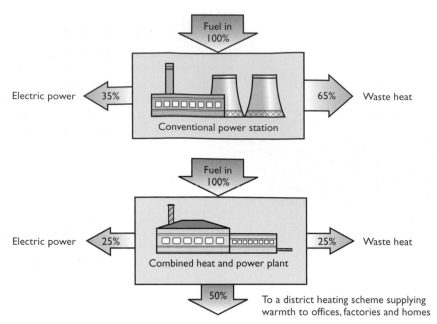

Figure 6.2 Efficiencies of conventional and combined heat power stations.

Renewable energy sources

◆ The Sun provides our planet with the energy required to drive the weather systems of the world and to allow plants to grow.
◆ The tides are a consequence of the gravitational pull of the Sun and the Moon.
◆ The Earth has a very hot interior which in some parts of the world comes close to the surface.

All of these offer alternative ways of obtaining the energy that we need. Scientists and engineers are slowly developing technologies to extract useful amounts of energy from them.

Solar power

The heat in sunlight (even on a dull Scottish day) is sufficient to heat water in panels (called solar panels) on the roofs of houses and so provide heating for the house. Sunlight can be used directly to generate electricity using **photocells** similar to those found in solar-powered calculators (figure 6.3).

Solar power is a renewable and clean source of energy. It is, however, difficult and expensive to convert large amounts into useful forms such as electrical energy, and this can only take place during daylight hours.

Wind power

The kinetic energy of the wind can be used to generate electricity using modern wind turbines. These turbines are normally situated in a windy location, such as a hilltop, in what are called wind farms. There are a large number of commercial **wind farms** in Britain (figure 6.4).

The wind is a renewable and clean source of energy. Conditions for wind energy in Britain are favourable since the prevailing winds are strong, especially in winter when energy demand is at its greatest. However, the wind cannot always be relied on to blow and so it is difficult to maintain a constant supply of electricity. In addition, wind generators can be unsightly and, since the best locations are generally in areas of great natural beauty, there have been environmental objections.

Figure 6.3 A house fitted with solar-powered heating panels.

Figure 6.4 Wind turbines grouped to form a wind farm.

Figure 6.5 A hydroelectric power station.

Hydroelectric power

The gravitational potential energy of water in a dam is used to generate electrical energy. It is widely used in the hilly parts of Scotland where there is ample rainfall (figure 6.5).

Hydroelectric power is a renewable, reliable and clean source of energy. However, there are environmental objections to the building of dams as they flood large areas of land and there is a risk (minimal) of a dam bursting.

Wave power

Winds cause waves to form on the sea. The energy in the waves can be extracted using a number of devices. Wave energy is more reliable than the wind, since the waves continue long after the wind which produced them has died away. However, there are large difficulties in the construction, maintenance and energy conversion of these devices and it will be some years before large-scale wave energy devices are available in significant numbers.

Tidal power

Some estuaries, such as that of the River Severn, provide suitable locations for tidal barrages. The principle is generally similar to hydroelectric power. Large gates are opened during the incoming tide allowing water to pass until high tide when they are closed. On the outgoing tide the potential energy of the trapped water drives turbines, allowing electricity to be generated. There are significant environmental objections to such schemes.

Geothermal power

The heat inside the Earth can be extracted when water is pumped into a bore hole. When the water boils, the steam formed can be used to drive turbines and generate electricity. Geothermal energy is a renewable and clean source of energy, but it is dependent on suitable sites, and the extraction of geothermal energy can present considerable technical difficulties. The heat that can be extracted from an area inside the Earth will decrease over a number of years. However, the heat will be slowly renewed if extraction is stopped.

Biomass (plants)

The most common way of extracting energy from biomass is to burn wood. Properly looked after forests, provide a sustainable source of energy. Alcohol distilled from plants is used in some countries to fuel cars (instead of petrol). Biomass is a renewable source of energy, but growth is too slow to provide large amounts as a fuel.

Nuclear power

Even in small quantities, uranium can produce large amounts of electrical energy. Known reserves of uranium will last a very long time, but are finite. The radioactive waste produced can be extremely dangerous and needs very careful handling and storage.

Table 6.1 summarises the renewable and non-renewable energy sources.

Renewable energy sources	Non-renewable (finite) energy sources
Solar	Coal
Wind	Oil
Hydroelectric	Gas
Wave	Nuclear
Tidal	–
Geothermal	–
Biomass (plants)	–

Table 6.1 Renewable and non-renewable energy sources.

Energy units

Different industries use different energy units, e.g. the electrical industry uses kilowatt hours; the food industry uses kilojoules. The unit used in physics for energy is the **joule** (J) (table 6.2).

Unit	Abbreviation	Number of joules
1 kilojoule	1 kJ	10^3 J = 1000 J
1 megajoule	1 MJ	10^6 J = 1 000 000 J
1 gigajoule	1 GJ	10^9 J = 1 000 000 000 J
1 kilowatt hour	1 kWh	3.6×10^6 J = 3 600 000 J

Table 6.2 Different energy units.

Section 6.1 Summary

◆ The fossil fuels coal, oil and gas are at present our main sources of energy.
◆ The reserves of the fossil fuels will not last for ever, i.e. they are finite.
◆ Examples of renewable energy sources are solar, wind, hydroelectric, wave, tidal, geothermal and biomass.
◆ Renewable energy sources have advantages and disadvantages, e.g. solar energy is non-polluting but is difficult and expensive to produce in large amounts.

---

1   The main sources of Britain's energy are the fossil fuels.

(a) Name these fossil fuels.
(b) Why are they called 'fossil fuels'?

2   Give **one** way in which energy could be conserved in:

(a) the home; (b) transport.

3   List the following sources of energy under the headings 'renewable' and 'non-renewable'.

Biomass, coal, gas, oil, solar, water, wind.

4   Explain the difference between renewable and non-renewable energy sources.

5   Give one advantage and one disadvantage of the following sources of energy:

(a) wind; (b) solar; (c) water.

6   A wind farm consists of 30 large wind generators. Each generator can produce 1.5 MW of electrical power.

(a) Suggest a suitable location for the wind farm.
(b) How much electrical power is the wind farm capable of generating?
(c) The wind farm produces an average 35 MW of electrical power for 20 hours. How many kWh of electricity is produced in the 20 hours?

7   The output power from a wave generator is on average 10 kW per metre. An island requires 20 MW of electrical power. Calculate the total length of wave generators required to supply power to the island.

Fossil-fuelled and nuclear power stations

Most of Britain's electrical energy is generated in large power stations using coal, gas, oil or nuclear fuel. Figure 6.6 shows a conventional (fossil-fuelled) power station. The chemical energy of the coal, natural gas or oil is changed into heat when the fuel is burned.

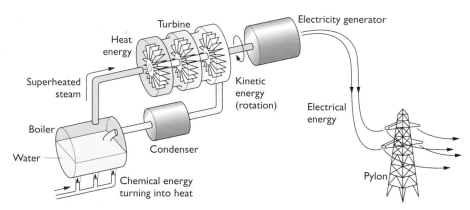

Figure 6.6 Layout of a fossil-fuelled power station.

The fuel of a nuclear power station is the uranium-235 nucleus. To obtain energy from these nuclei, they are bombarded with neutrons. **Neutrons** are uncharged particles found in the nucleus of an atom. When a neutron strikes and is absorbed by a uranium-235 nucleus, the nucleus becomes unstable and splits into two pieces. The splitting of a uranium-235 nucleus is called **nuclear fission** and the pieces produced are called **fission fragments** (figure 6.7(a)). A large amount of (heat) energy is released during the fission of a uranium-235 nucleus together with two further neutrons. These neutrons can cause other fissions of uranium-235 nuclei and these, in turn, release further neutrons which can cause further fissions. When a continuous reaction of fissions occurs this is a called a **chain reaction** (figure 6.7(b)).

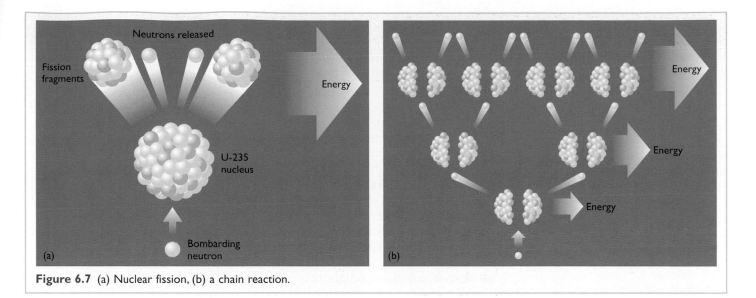

Figure 6.7 (a) Nuclear fission, (b) a chain reaction.

In a nuclear power station, a reactor replaces the boiler used in a conventional power station. The nuclear energy of the uranium-235 fuel is changed to heat during the fission process. Carbon dioxide gas then carries heat produced by the fission of uranium-235 nuclei in the reactor to a heat exchanger. The heat exchanger transfers heat from the carbon dioxide to water, which turns into steam. The steam is used to turn turbines, just as it would in a conventional power station. Figure 6.8 shows part of a nuclear power station. Control rods are used to control the number of fissions occurring every second. The control rods are made from a substance which absorbs neutrons, such as boron. By lowering the rods into the reactor, neutrons are absorbed and the number of fissions produced every second is reduced; by raising them the number of fissions produced every second is increased. As slow moving neutrons are more likely to cause fission to occur, a moderator, such as graphite (carbon) is used to slow the fast moving neutrons produced during the fission process.

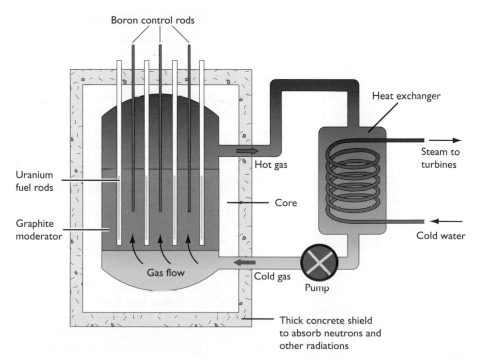

Figure 6.8 Part of a nuclear power station.

A nuclear power station requires much less fuel than the equivalent coal fired power station.

This is because 1 kg of coal produces about 30 MJ of energy when burnt, but 1 kg of uranium fuel produces 5 000 000 MJ of energy during the fission process.

Nuclear power stations do, however, produce radioactive waste which can be very dangerous to us and the environment. This waste has to be safely stored for many years.

Hydroelectric power stations

About two per cent of Britain's electrical energy is obtained from hydroelectric power stations. Water in a high level dam or reservoir has gravitational potential energy. Hydroelectric power generation makes use of this energy. The water is allowed to flow in pipes down a steep hill, to turn turbines which drive generators to produce electricity. Figure 6.9 shows a diagram of a hydroelectric power station.

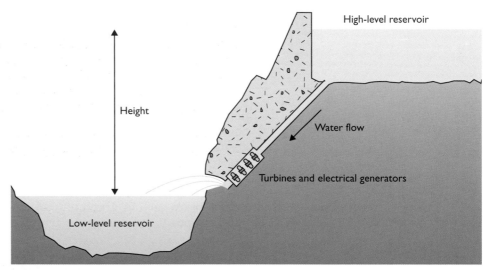

Figure 6.9 A hydroelectric power station.

The amount of electrical energy produced is dependent on the mass and height of the water above the generating station, i.e. the water's change in gravitational potential energy.

$$\text{Change in gravitational potential energy, } E_p = mgh$$

Pumped hydroelectric power stations

Electrical energy in large quantities cannot be stored and so it must be used as it is generated. However, demand for electrical energy changes with the time of day and the season of the year. Figure 6.10 shows how the demand for electrical power changes on a typical winter day.

In a hydroelectric station with pumped storage, electrical energy generated at 'off-peak' periods of the day or night is used to pump water back up from a low-level reservoir to a high-level reservoir. At 'peak' periods of the day, the gravitational potential energy of the water in the high-level reservoir is converted back into electrical energy for the National Grid. Although only three kWh (kilowatt hours) of electrical energy are recovered for every four supplied, i.e. the system is 75 per cent efficient,

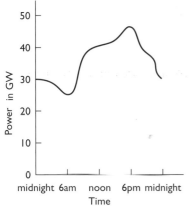

Figure 6.10 Electrical power demand varies during a winter day.

CHAPTER SIX **Energy Matters**

pumped-storage schemes are an economical way of meeting peak demand, and they help to increase the overall efficiency of the electrical energy supply system.

Efficiency

The efficiency of a power station (or any other machine) is given by:

$$\text{Efficiency} = \frac{\text{useful energy output}}{\text{total energy input}} \times 100\%$$

As the energy output from a power station or any other machine is always less than the energy input, the efficiency is always less than 100 per cent. The useful energy we get from a machine is never as great as the energy we put in. This is because some of the energy is transferred into other forms such as heat and perhaps sound. We say that some of the energy put in to the machine has been 'lost'. However, the total amount of all the energies remains the same, i.e. the total energy is **conserved**. Sometimes it is easier to look at the energy being used every second (the power). In this case 'power' replaces 'energy' in the equation.

Example
A hydroelectric power station is able to convert 75 per cent of the water's gravitational potential energy into electricity. Water in the hydroelectric power station falls through a vertical height of 50 m. Water passes through the turbine at the rate of 2000 kg every second.
(a) Calculate the loss in gravitational potential of the water in one second.
(b) What is the power output of the station?
(c) The efficiency of the power station is 75 per cent. What happened to the other 25 per cent?

Solution

(a) In one second, loss in gravitational $E_\text{p} = mgh$
$$= 2000 \times 10 \times 50 = 1 \times 10^6\,\text{J}$$

(b)
$$\text{Efficiency} = \frac{\text{useful energy output}}{\text{total energy input}} \times 100\%$$

$$75\% = \frac{\text{useful energy output}}{1 \times 10^6} \times 100\%$$

$$\text{useful energy output} = \frac{75\%}{100\%} \times 1 \times 10^6 = 7.5 \times 10^5\,\text{J},$$

i.e. energy output per second $= 7.5 \times 10^5\,\text{J}$

Hence power output $= 7.5 \times 10^5\,\text{W}$

(c) Twenty-five per cent of the energy input was changed into heat and sound in the power station.

Useful and useless energy

Energy cannot be created or destroyed, it is simply changed from one form into another. There is always the same amount of energy, although it may be in a number of different forms after the energy change has taken place. However, less of the energy after the change is 'useful' – the rest being in a form or forms that cannot be easily used. Energies, such as heat, light and sound, which cannot easily be used again are said to be degraded. The following provides an illustration of degraded energy.

One joule of electrical energy can produce one joule of heat in a resistor, but this heat cannot reproduce one joule of electrical energy. The electrical energy produced would be less than one joule as some heat would be 'lost' to the atmosphere and could not be recovered.

Section 6.2 Summary

- Thermal power stations change the chemical energy of the fuel (coal, oil, gas) into electrical energy (chemical to heat to kinetic to electrical).
- A nuclear power station changes the nuclear energy of the uranium fuel into electrical energy (nuclear to heat to kinetic to electrical).
- A hydroelectric power station changes the gravitational potential energy of water behind a high dam into electrical energy (gravitational potential to kinetic to electrical).
- A nuclear power station produces radioactive waste.
- Nuclear fission can occur when a uranium-235 nucleus is struck by and absorbs a neutron. The uranium-235 nucleus becomes unstable and splits into two large fragments and some neutrons. The 'extra' neutrons can go on to produce further fissions.
- A chain reaction is when at least one neutron from each fission goes on to produce a further fission.
- Energy is degraded during any energy change. This means that there is less useful energy after the change than was available before the change took place.

End of Section Questions

1 An electric motor is used to raise a mass from the floor. During the lifting, the mass gained 12 J of gravitational potential energy. The electrical energy supplied to the motor during the lifting process was 36 J. Calculate the efficiency of the system during the lifting operation.

2 In a small-scale hydroelectric scheme, water falls through a vertical height of 70 m. On average, 750 kg of water passes through the turbine every five seconds.
   (a) Calculate the loss of gravitational potential energy by the water every five seconds.
   (b) Find the maximum power output from the hydroelectric scheme.

3 A small wind turbine is used to provide electrical energy for a roadside sign. The wind turbine converts 30 per cent of the wind's power into electrical power. During part of the day the constant electrical power output of the turbine is 10 W. Calculate the power input to the wind turbine at this time.

4 The power input to a gas-fired power station is 1800 MW. The power station produces 860 MW of electrical power. Calculate the efficiency of the power station at producing electricity.

At the end of this section you should be able to:

1 Identify circumstances in which a voltage will be induced in a conductor.
2 Identify on a given diagram the main parts of an a.c. generator.
3 State that transformers are used to change the magnitude of an a.c. voltage.
4 Describe the structure of a transformer.
5 Carry out calculations involving the relationship between V_s, V_p, N_s, N_p.
6 State that high voltages are used in the transmission of electricity to reduce power loss.
7 Describe qualitatively, the transmission of electrical energy by the National Grid system.

8 Explain from a diagram how an a.c. generator works.
9 State the main differences between a full-size generator and a simple working model.
10 State the factors which affect the size of the induced voltage, i.e. field strength, number of turns on the coil, relative speed of magnet and coil.
11 Explain why a transformer is not 100% efficient.
12 Carry out calculations on transformers involving input and output voltages, turns ratio, primary and secondary currents and efficiency.
13 Carry out calculations involving power loss in transmission lines.

Generating electricity

In Section 2.6 you saw how movement (or a force) was produced when a current-carrying wire was placed in a magnetic field (see page 58). However, the opposite effect is possible. Figure 6.11 shows a coil of wire connected to the input of an oscilloscope. The wire is being moved through a magnetic field. When the coil is moved up or down through the magnetic field we find that a voltage is produced across the ends of the coil.

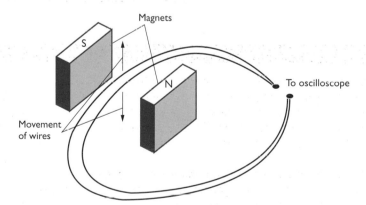

Figure 6.11 Generating electricity by moving a wire through a magnetic field.

The size of this voltage is dependent on:
◆ The number of turns of wire on the coil – the greater the number of turns, the greater the voltage produced.
◆ The strength of the magnetic field – the stronger the magnetic field, the greater the voltage produced.
◆ The speed of movement – the faster the coil is moved up or down through the magnetic field, the greater the voltage produced.

A simple generator

In Section 2.6 you looked at how an electric motor changed electrical energy into kinetic energy. The motor, however, can be made to work in reverse, i.e. to change kinetic energy into electrical energy. When the motor is used in this way it is acting as a **dynamo** or **generator**. Figure 6.12 shows an 'electric motor' being used to generate electricity.

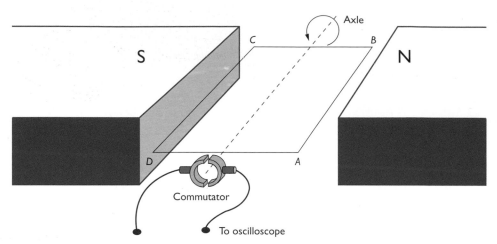

Figure 6.12 A dynamo is an electric motor in 'reverse'.

Generation of a.c.

Alternating current (a.c.) can be produced by using slip rings which rotate with the coil, as shown in figure 6.13(a). The slip rings ensure one brush is always connected to one side of the coil whether it is moving up or down through the magnetic field. Figure 6.13(b) shows the output from the generator and figure 6.13(c) shows the coil at various positions as it makes one revolution. As the coil rotates, the right-hand side, AB, and the left-hand side, CD, of the coil move through the magnetic field. The magnetic field through the coil changes and a voltage is produced. When the coil is in the vertical position, there is no change in magnetic field through the coil and so there is no voltage. When the coil moves through the vertical position, the voltage in the coil is now in the opposite direction as the direction of motion has changed, since AB is now moving down and CD is moving up. The direction of the voltage changes each time the coil rotates through half a revolution, i.e. alternating current is produced. This type of generator is often called an **alternator**.

A commercial alternator works in a slightly different way to the a.c. generator discussed above. Figure 6.14 shows the major components of a commercial alternator. Instead of having a rotating coil and a stationary magnet, alternators have rotating electromagnets (called the **rotor** or **field coils**) and stationary coils (called the **stator**). There are two main advantages to this type of design: electromagnets can provide a stronger magnetic field than permanent magnets of the same size; and there are no moving parts needed to collect the large electrical current generated.

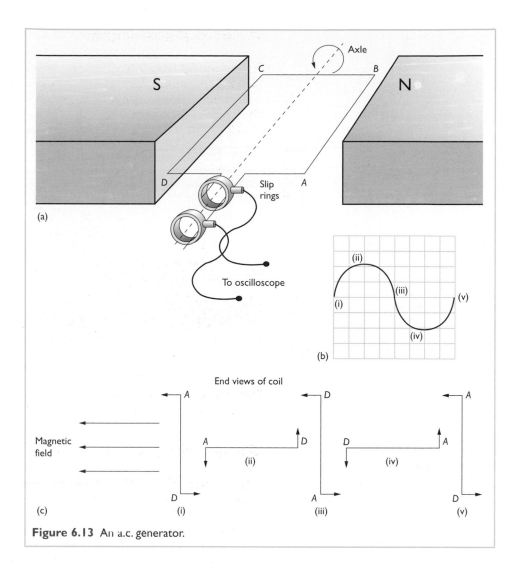

(a)

(b)

End views of coil

(c)

Figure 6.13 An a.c. generator.

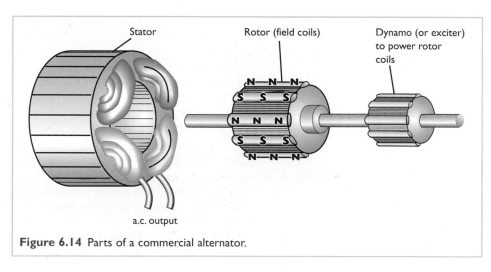

Figure 6.14 Parts of a commercial alternator.

Figure 6.15 shows the layout of a conventional power station.

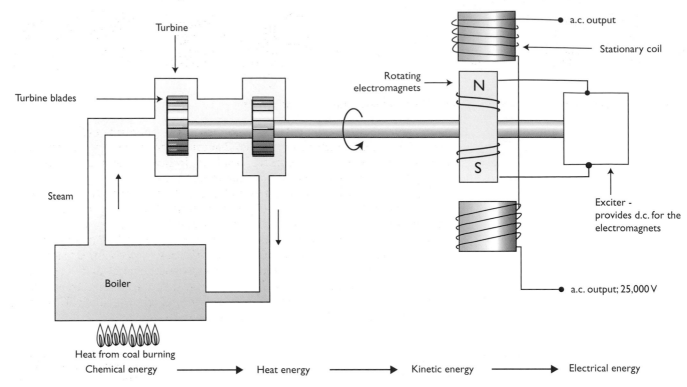

Chemical energy   ⟶   Heat energy   ⟶   Kinetic energy   ⟶   Electrical energy

Figure 6.15 A power station.

Transmitting electrical energy

Figure 6.16 shows how electrical energy, generated at a power station, is distributed through the National Grid transmission system to our homes.

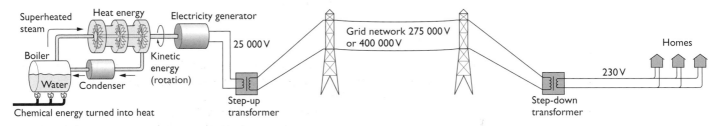

Figure 6.16 Power transmission and distribution.

The generator at the power station produces a voltage of 25 000 volts. For efficient transmission over long distances this voltage is increased by a (step-up) transformer, to 275 000 volts (or 400 000 volts for the National Grid system). At the end of the transmission lines a (step-down) transformer reduces this voltage for distribution to consumers.

Model power line

Figure 6.17 shows an experiment to demonstrate why transformers are required in the transmission of electrical energy. In figure 6.17(a) electrical energy is transmitted directly along the transmission lines. Lamp Y is less brightly lit than lamp X. This is due to some of the electrical energy being changed to heat in the transmission lines. The amount of energy changed to heat every second (the power loss) is given by $P = I^2R$, where I is the current through the transmission line and R is the total resistance of the transmission lines.

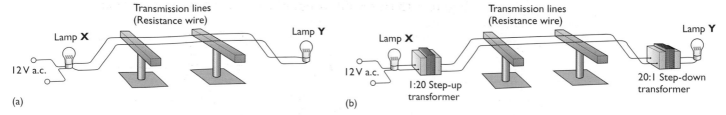

(a)

(b)

Figure 6.17 Figures (a) and (b) show power transmission without and with transformers.

In figure 6.17(b) the voltage is increased (stepped up) before transmission, and reduced (stepped down) at the far end of the transmission lines. In this case lamp Y is much brighter, indicating that much less heat energy has been produced in the transmission lines.

When electrical power is passed along transmission lines, it is important to keep the power loss as low as possible. Since the power loss from the transmission lines $= I^2R$, the current through the transmission lines, and the resistance of the transmission lines, should both be as low as possible.

In practice there is a limit to how low the resistance of the transmission lines can be economically made. However, transformers make it possible to reduce the size of the current through the transmission lines by increasing the voltage. Transformers are therefore essential when electrical energy is to be transmitted over large distances (figure 6.18).

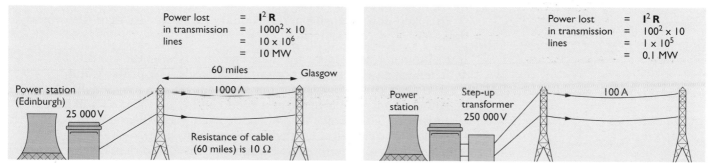

Power lost in transmission lines	= I^2R
	= $1000^2 \times 10$
	= 10×10^6
	= 10 MW

60 miles — Glasgow

1000 A

Power station (Edinburgh)

25 000 V

Resistance of cable (60 miles) is 10 Ω

Power lost in transmission lines	= I^2R
	= $100^2 \times 10$
	= 1×10^5
	= 0.1 MW

Power station

Step-up transformer 250 000 V

100 A

Figure 6.18 Power loss in transmission lines can be greatly reduced by the use of transformers.

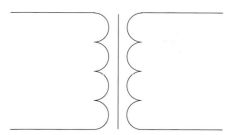

Figure 6.19 Circuit symbol for a transformer.

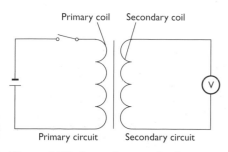

Figure 6.20 A transformer circuit.

The transformer

A transformer consists of two separate coils of wire wound on the same iron core. The circuit symbol for a transformer is shown in figure 6.19. The straight line between the coils is the iron core.

Figure 6.20 shows a transformer connected to a switch and a battery. When the switch is closed, a current passes through the primary coil. This current produces a magnetic field in the primary coil. This magnetic field rapidly builds up through both sets of coils, since they are joined by the iron core. This gives a changing magnetic field in the secondary coil and so a voltage is produced.

When the current in the primary circuit is steady, there is no change in the magnetic field and no voltage is produced in the secondary circuit. However, when the switch is opened, the primary current is switched off and the magnetic field around both the primary and secondary coils rapidly collapses (disappears). This changing magnetic field through the secondary coil results in a voltage being produced but in the opposite direction. Removing the iron core through the coils decreases the effect, since the iron core concentrates the magnetic field through the coils. With a d.c. supply connected to the primary coil, a changing magnetic field can only be obtained by opening and closing the switch. However, a more practical way of obtaining a changing magnetic field is to connect the primary coil to an a.c. supply. The a.c. current is always changing in size and direction. Transformers, therefore only work on a.c.

A transformer is used to investigate the relationship between the alternating voltages at the primary and secondary coils and the number of turns on the primary and secondary coils.

Primary turns	Secondary turns (N_s)	$\dfrac{N_s}{N_p}$	Primary voltage (V_p) V	Secondary voltage (V_s) V	$\dfrac{V_s}{V_p}$
125	500	4	2	8	4
500	125	0.25	2	0.5	0.25
125	625	5	2	10	5
500	500	1	2	2	1
625	125	0.2	2	0.4	0.2

Table 6.3

From the results in table 6.3 we see that:

$$\frac{N_s}{N_p} = \frac{V_s}{V_p}$$

In a step-up transformer: secondary turns are more than the primary turns, i.e. $N_s > N_p$ thus $V_s > V_p$.

In a step-down transformer: secondary turns are less than the primary turns, i.e. $N_s < N_p$ thus $V_s < V_p$.

In real transformers, there are some energy losses. These are due to:
◆ The heating effect of the current in the coils. The primary and secondary coils of a transformer are made up of a long length of wire. Although the wire is made of a good conductor, the coils still have a resistance. This means that when a current passes through the coils, some of the electrical energy is changed into heat.
◆ The iron core is continually being magnetised and demagnetised, and this results in heat being produced.
◆ The transformer vibrates as a result of the magnetising and demagnetising of the iron core and so sound is produced.
◆ Some of the magnetic field produced by the primary 'leaks' from the iron core and so does not pass through the secondary coil. This results in a smaller voltage being produced at the secondary coil.

Since the energy losses in transformers are normally very small (about 5 per cent to heat, sound and leakage) it is convenient when doing problems to assume that the transformer is 100 per cent efficient, i.e. an ideal transformer.

For an **ideal** transformer, all the power produced at the primary is transferred to the secondary. Therefore:

input power = output power

$$I_p \times V_p = I_s \times V_s$$

$$\text{i.e. } \frac{V_s}{V_p} = \frac{I_p}{I_s}$$

Hence:

$$\frac{V_s}{V_p} = \frac{N_s}{N_p} = \frac{I_p}{I_s}$$

Example

An ideal transformer steps down the mains voltage of 230 V a.c. to 3 V a.c. The secondary coil of the transformer has 300 turns. Calculate the number of turns on the primary coil of the transformer.

Solution

$$\frac{N_p}{N_s} = \frac{V_p}{V_s}$$

$$\frac{N_p}{300} = \frac{230}{3}$$

$$N_p = \frac{230}{3} \times 300 = 23\,000 \text{ turns}$$

Example

An ideal transformer has 4000 primary turns and 100 secondary turns. The current in the primary coil of the transformer is 10 mA. Calculate the current in the secondary coil of the transformer.

Solution

$$\frac{I_s}{I_p} = \frac{N_p}{N_s}$$

$$\frac{I_s}{0.01} = \frac{4000}{100}$$

$$I_s = 0.01 \times 40$$

$$I_s = 0.4 \text{ A}$$

Example

A transformer steps down the mains voltage of 230 V a.c. to 12 V a.c. The current in the primary coil of the transformer is 0.1 A. The transformer is 95% efficient.
(a) Calculate the power input to the transformer.
(b) What is the power output from the transformer?
(c) Calculate the current in the secondary coil.

Solution

(a) **Power input to transformer** $= IV = 230 \times 0.1 = 23$ **W**

(b)
$$\text{Efficiency} = \frac{\text{useful power output}}{\text{total power input}} \times 100\%$$

$$95\% = \frac{\text{power output}}{23} \times 100\%$$

$$\text{power output} = \frac{95\%}{100\%} \times 23 = 21.9 \text{ W}$$

(c)
$$\text{Power output} = IV$$
$$21.9 = 12 \times I$$

$$I = \frac{21.9}{12} = 1.8 \text{ A}$$

Section 6.3 Summary

◆ A voltage can be produced in a coil of wire when the magnetic field near the coil changes.

◆ The size of the induced voltage can be increased by increasing the strength of the magnetic field, increasing the number of turns on the coil and increasing the speed of the coils as they pass through the magnetic field.

◆ A transformer consists of two separate coils of wire wound on an iron core.

◆ Transformers are used to change the size of an a.c. voltage.

◆ For a transformer $$\frac{V_p}{V_s} = \frac{N_p}{N_s}$$

◆ For an ideal transformer, i.e. 100% efficient $$\frac{V_p}{V_s} = \frac{N_p}{N_s} = \frac{I_s}{I_p}$$

◆ High voltages are used in the transmission of electrical power as this reduces the current that the transmission lines carry and so the power loss ($= I^2 R$) in the transmission lines is reduced.

◆ Electrical power is distributed by the National Grid system – electrical power produced at the power station at 25 000 V is stepped-up by a transformer to 400 000 V for efficient transmission along the transmission lines. At the other end of the transmission lines a step-down transformer reduces the voltage to 230 V for use in our homes.

End of Section Questions

1 The figure below shows a loop of wire connected to a voltmeter.

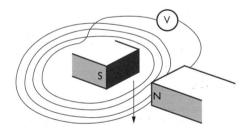

Figure 6.3Q1

When the loop of wire is moved between the poles of the magnet (as shown) a reading is obtained on the voltmeter. State three changes that could be made to increase the reading on the voltmeter.

2 A simplified diagram of an a.c. generator is shown in the figure below.

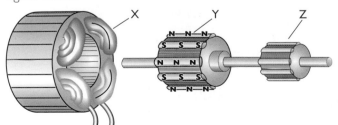

Figure 6.3Q2

The main parts of the generator are: dynamo (exciter), rotor (field coils), stator.
Identify, using the terms above, the parts labelled X, Y and Z.

3 A lamp is connected to the secondary coil of the transformer as shown in the figure below.

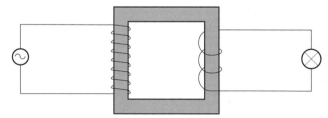

Figure 6.3Q3

An a.c. source is applied to the primary coil of the transformer.
(a) The lamp is brightly lit. Is the current in the secondary coil a.c. or d.c.?
(b) The a.c. source is replaced with a d.c. source. Explain the effect this will have on the lamp.

4 Calculate the unknown quantity or quantities for the ideal transformers shown below.

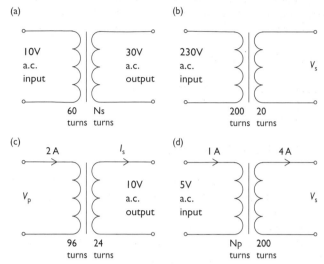

Figure 6.3Q4

5 A 230 V a.c. supply is connected to the 1000 turn primary coil of an ideal transformer. A 460 W electric drill is connected across the 500 turn secondary coil of the transformer.

(a) What is the voltage across the drill?
(b) Calculate the current passing through the drill when it is operating at its correct power rating.
(c) Calculate the current in the primary coil of the transformer when the drill is operating.

6 A transformer is not 100 per cent efficient, but wastes some energy. Give two reasons why some of the input energy to the transformer is wasted.

7 A step-up transformer is used to supply current to a 20 Ω resistor from a 2 V a.c. supply. The reading on a voltmeter connected across the resistor is 8 V as shown in the figure below.

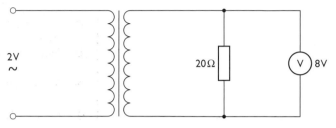

Figure 6.3Q7

(a) Calculate the current through the resistor.
(b) How much power is used up in the resistor?
(c) The current in the primary circuit is 2 A. How much power is supplied to the transformer?
(d) What is the efficiency of the transformer?

8 A model transmission system is shown in the figure below. The transmission lines have a total resistance of 5 Ω. Both transformers may be assumed to be ideal.

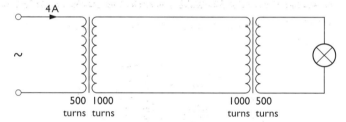

Figure 6.3Q8

(a) Calculate the current in the transmission lines.
(b) Calculate the power loss from the transmission lines.

At the end of this section you should be able to:

1　Use the following terms correctly in context: temperature; heat; Celsius.
2　Describe two ways of reducing heat loss in the home due to: conduction; convection; radiation.
3　State that heat loss in a given time depends upon the temperature difference between the inside and the outside of the house.
4　State that the same mass of different materials requires different quantities of energy to raise their temperature of unit mass by one degree.
5　Carry out calculations based on practical applications involving heat, mass, specific heat capacity and temperature change.
6　Give examples of applications which involve a change of state, e.g. refrigerator or picnic box cooler.
7　Use the following terms correctly in context: specific heat capacity; change of state; latent heat of fusion; latent heat of vaporisation.
8　State that a change of state does not involve a change of temperature.
9　State that energy is gained or lost by a substance when its state is changed.
10　Use the principle of conservation of energy to carry out calculations on energy transformations which involve temperature change, e.g. $IVt = E_h = cm\Delta T$.
11　Carry out calculations involving specific latent heat.

Heat and temperature

The terms heat and temperature are very often confused. Heat, just like light and sound, is a form of energy and is measured in joules (J). Temperature is a measure of how hot a substance is, and is measured in degrees Celsius (°C). The discussion below illustrates the difference between them.

Two pupils, Jack and Jean, each heat water in identical kettles. Jack heats 0.3 kg of water for one minute. Jean heats 0.6 kg of water for one minute. During this time the kettles both supplied the same amount of heat to the water, but the temperature of Jack's water was higher than that of Jean's.

Heat transfer

Heat energy always moves from a hotter substance (high temperature) to the cooler surroundings (low temperature). There are three possible ways in which it can move: **conduction**, **convection** and **radiation**.

Conduction

◆ Heat is transferred through a solid material by conduction. The particles making up the solid cannot change their position, but pass the heat from particle to particle.
◆ Heat moves from a high temperature to a low temperature.
◆ Materials which allow heat to move easily through them are called **conductors** while materials which do not allow heat to move through them easily are called **insulators**. Metals are the best conductors, while liquids and gases are good insulators (poor conductors).

Convection

Convection occurs in liquids and gases. Heat is transferred by the movement of the heated particles making up the liquid or gas. The heated fluid (liquid or gas) becomes less dense and rises, so cold fluid falls to take its place, i.e. convection currents are set up.

◆ Convection cannot take place in a solid.
◆ Many heat insulators contain trapped air, e.g. cotton wool, felt, woollen clothes. Convection cannot take place because the air is trapped and cannot move. Also, air is a non-metal, so it is a poor conductor of heat.

Radiation

◆ Radiation travels in straight lines until absorbed by an object. It can travel through a vacuum (the Earth is heated by radiation from the Sun).
◆ All hot materials radiate heat energy.

Heat loss and temperature difference

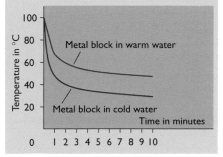

Figure 6.21 Graphs of temperature against time for both metal blocks.

David and Mary decided to investigate how heat loss depended on the temperature of the surroundings. They heated two identical metal blocks to 100°C. One block was placed into cold water and the other into warm water. The temperature of each block was measured every minute for ten minutes. The results were displayed on a graph, as shown in figure 6.21. From their investigation they concluded that the metal block in the cold water lost more heat energy than the block in the warm water.

The amount of heat energy lost every second depends on the difference in temperature between an object and its surroundings. The greater the difference in temperature, the greater the loss of heat energy in one second. The smaller the difference in temperature, the smaller the loss of heat energy in one second.

Heat loss from houses

Figure 6.22 shows the main heat losses from a house. Such heat losses can be reduced by insulation.

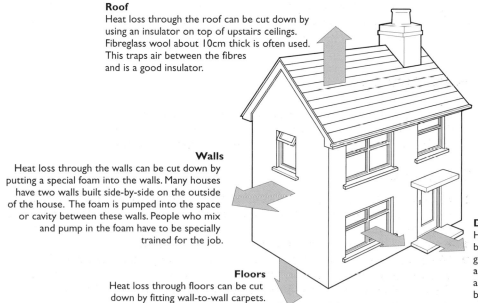

Roof
Heat loss through the roof can be cut down by using an insulator on top of upstairs ceilings. Fibreglass wool about 10cm thick is often used. This traps air between the fibres and is a good insulator.

Walls
Heat loss through the walls can be cut down by putting a special foam into the walls. Many houses have two walls built side-by-side on the outside of the house. The foam is pumped into the space or cavity between these walls. People who mix and pump in the foam have to be specially trained for the job.

Floors
Heat loss through floors can be cut down by fitting wall-to-wall carpets.

Doors and windows
Heat loss through the windows can be cut down by fitting an extra pane of glass. This is called double glazing. The two panes are a few millimetres apart and trap a thin layer of air. Curtains (when closed) also help to cut heat loss. Draughts can be stopped by sealing doors and windows.

Figure 6.22 Main heat losses from a house.

In most houses, hot water is stored in a large copper tank. If the tank is not insulated, a great deal of heat will be lost by convection. This is because the temperature of the hot water is usually about 60°C, while the temperature of the surrounding air is about 18°C.

Draughts are a very obvious and uncomfortable sign of a badly insulated house. Wherever you can feel cold air coming in, warm air is rushing out.

More heat energy is lost through the walls of an average uninsulated house than by any other route. However, effective wall insulation can reduce this heat loss by up to two-thirds. The floors of your house can also be insulated and the windows double- or triple-glazed. Insulating the loft of your house can cut as much as 20 per cent off your energy bill.

Changing temperature

When you want to make a hot drink, you put some water in the kettle and switch on. The water will get hot as a result of the heat energy supplied by the kettle. However, the amount of heat energy required to warm the water depends on:

◆ the temperature rise – more energy is needed for a larger temperature rise;
◆ the mass of the water – more energy is needed for a greater mass of water.

Example
20 900 J of energy is required to increase the temperature of 0.5 kg of water by 10°C.
(a) How much energy is required to increase the temperature of 0.5 kg of water by 20°C?
(b) How much energy is required to increase the temperature of 1 kg of water by 40°C?

Solution
(a) It takes 20 900 J to heat up 0.5 kg of water by 10°C, so it will take twice as much energy to heat 0.5 kg by 20°C.
Therefore, it takes 41 800 J to heat up 0.5 kg of water by 20°C.
(b) It takes 20 900 J to heat up 0.5 kg of water by 10°C.
So it takes 41 800 J to heat up 1 kg of water by 10°C (twice the mass so twice the energy).
Therefore, it takes 167 200 J to heat up 1 kg of water by 40°C (four times the change in temperature so four times the energy).

Specific heat capacity

So far we have only considered heating the same substance, namely water. However, if you were to heat equal masses of water and copper with the same quantity of heat energy, then you would find that the copper would have a much higher rise in temperature. The heat energy needed to change the temperature of a substance depends on:

◆ the change in temperature of the material (ΔT);
◆ the mass of the material (m);
◆ the type of material (specific heat capacity of the material).

The **specific heat capacity** of a substance is the amount of energy required to change the temperature of 1 kg of a substance by 1°C. The units of specific heat capacity are joules per kilogram per degree Celsius (J/kg°C). Water has a specific heat capacity of 4180 joules per kilogram per degree Celsius (4180 J/kg°C). This means that it takes 4180 joules of heat energy to change the temperature of 1 kg of water by 1°C.

It would take 16 720 J to heat up 4 kg of water by 1°C (four times the mass so four times the energy) and 83 600 J to heat up 4 kg of water by 5°C (five times the change in temperature so five times the energy). This can be put in the form of an equation:

Energy required to change temperature of the substance = specific heat capacity × mass × change in temperature

$$E_h = c \times m \times \Delta T$$

where E_h = energy needed to change the temperature of the substance (J)

c = specific heat capacity of substance (J/kg°C)

m = mass of substance (kg)

ΔT = change in temperature of substance (°C)

Table 6.4 shows the specific heat capacities of some substances.

Material	Specific heat capacity in J/kg °C	Material	Specific heat capacity in J/kg °C
Water	4180	Sea water	3900
Lead	128	Ice	2100
Aluminium	902	Concrete	800
Alcohol	2350	Glass	500
Steel	500	Copper	386

Table 6.4 Specific heat capacities.

Example
How much heat energy is required to increase the temperature of 0.2 kg of water from 20°C to 50°C?
Solution

$$E_h = c\, m\, \Delta T$$

$$E_h = 4180 \times 0.2 \times (50 - 20)$$

$$E_h = 4180 \times 0.2 \times 30$$

$$E_h = 25\,080\,J$$

Example
A night-storage heater contains 70 kg of concrete bricks. How much heat energy is released by the bricks when they cool from 90°C to 20°C?
Solution

$$E_h = c\, m\, \Delta T$$

$$E_h = 800 \times 70 \times (90 - 20)$$

$$E_h = 800 \times 70 \times 70$$

$$E_h = 3\,920\,000\,J$$

Heat problems

Although energy can be changed from one form to another, the total amount remains unchanged. This is the **principle of conservation of energy** – an electric heater converts electrical energy into an equal amount of heat energy. Due to conduction, convection and radiation, some of the heat energy supplied will be transferred ('lost') to the surroundings. This means that the substance will absorb (take in) less energy than was supplied by the heater.

Energy supplied =
energy absorbed + energy transferred to the surroundings

In most heat problems it is assumed that no energy is transferred to the surroundings, so:

Energy supplied (by heater) =
energy absorbed (by the substance)

Example
A small immersion heater takes 10 minutes to raise the temperature of 0.25 kg of water from 16°C to 45°C. The water is heated in a well-insulated cup. Calculate the power rating of the heater.

Solution

$$E_h = c\, m\, \Delta T = 4180 \times 0.25 \times (45 - 16)$$

$$E_h = 4180 \times 0.25 \times 29$$

$$E_h = 30\,305\,\text{J}$$

$$P = \frac{E}{t} = \frac{30\,305}{10 \times 60} = 50.5 = 51\,\text{W}$$

Example
A deep-fat fryer is used to heat 750 g of cooking oil from 20°C to 140°C in a time of 180 s. The specific heat capacity of the cooking oil is 3000 J/kg°C.
(a) Calculate the power rating of the element of the deep-fat fryer.
(b) The element of the deep-fat fryer is connected to a 230 V mains supply. Find the current in the element.

Solution

(a)
$$E_h = c\, m\, \Delta T = 3000 \times 0.75 \times (140 - 20)$$

$$E_h = 3000 \times 0.75 \times 120 = 270\,000\,\text{J}$$

$$P = \frac{E}{t} = \frac{270\,000}{180} = 1500\,\text{W}$$

(b)
$$P = IV$$

$$1500 = I \times 230$$

$$I = \frac{1500}{230} = 6.5\,\text{A}$$

Specific latent heat

When cold water in a kettle is heated, the temperature of the water rises until the water starts to boil at 100°C. Further heating of the water no longer produces a rise in temperature of the water, but steam at 100°C is produced (i.e. the heat energy supplied by the kettle is now being used to change water at 100°C into steam at 100°C). The energy required to change 1 kg of a liquid at its boiling point into 1 kg of vapour is called the **specific latent heat of vaporisation**. The word **latent** means 'hidden' and refers to the fact that the temperature does not rise, and the heat energy supplied to it seems to have disappeared.

When ice, at its melting point of 0°C, is heated, it turns into water at 0°C. Heat energy is required to change the ice to water at constant temperature. The energy required to change 1 kg of a solid at its melting point into 1 kg of liquid is called the **specific latent heat of fusion**.

The three **states** of matter are solid, liquid and gas. Whenever a substance changes state, latent heat energy is required. When a substance changes from a solid to a liquid, or a liquid to a gas, energy is needed to break down the force (or bond) holding the particles together and to push the particles further apart.

Specific latent heat of fusion

The energy required to change 1 kg from solid at its melting point to liquid without a change in temperature.

Specific latent heat of vaporisation

The energy required to change 1 kg from liquid at its boiling point to gas without a change in temperature.

When a substance changes state from solid to liquid or liquid to gas, latent heat is absorbed (taken in). When a substance changes state from gas to liquid or liquid to solid, latent heat is released (given out).

The symbol for specific latent heat is l. Specific latent heat is measured in J/kg.

For a substance with a specific latent heat l:
- to change the state of 1 kg of the substance at constant temperature requires 1 J,
- to change the state of m kg of the substance at constant temperature requires $m \times 1$ J, i.e. $E_h = m\,l$.

Energy needed to change state = mass × specific latent heat

$$E_h = m\,l$$

where E_h = energy needed to change state (J),

m = mass which changed state (kg),

l = specific latent heat (J/kg)

Example

A 36 W immersion heater is used to bring 0.5 kg of a liquid to its boiling point. The heater is left on for a further 300 s. Find the mass of liquid boiled off in this time. The specific latent heat of vaporisation of the liquid is 200 000 J/kg.

Solution

$$P = \frac{E}{t}$$

$$36 = \frac{E}{300}$$

$$E = 36 \times 300 = 10\,800\,\text{J}$$

$$E_h = m\,l$$

$$10\,800 = m \times 200\,000$$

$$m = \frac{10\,800}{200\,000} = 0.054\,\text{kg}$$

Mass of liquid boiled off is 0.054 kg.

Specific latent heat of vaporisation of water

The specific latent heat of vaporisation of water can be measured using the apparatus shown in figure 6.23. The lid of the kettle is left off so that the automatic cutout does not work.

Mass of water changed to steam = 0.1 kg

Power rating of kettle = 2000 W

Time of supply = 116 s

Energy supplied by heater = $P \times t = 2000 \times 116 = 232\,000\,\text{J}$

Using $E_h = m\,l$

$$232\,000 = 0.11$$

$$l = 2\,320\,000\,\text{J/kg}$$

In practice, energy is lost to the surroundings, and so the above result is only an estimate. The accepted value for the specific latent heat of vaporisation of water is 2 260 000 J/kg.

The difference between the experimental result and the accepted value can be explained as:

◆ Not all of the energy supplied by the heater element is absorbed by the water. Some of it is 'lost' in heating up the air around the kettle. This means that more energy is required to be supplied to produce 0.1 kg of steam and so the value for *l* is too large.

◆ Some of the steam condensed on the cooler parts of the kettle and fell back in to be reheated – this results in more energy being supplied to produce 0.1 kg of steam and this makes the value for *l* too large.

Freezing food

Unless special precautions are taken, most fruit and vegetables that are frozen using a normal freezer are damaged. During the freezing process, the structure of the cells making up the food is damaged. When the food thaws, most of the water in the food is released and an unappetising 'mush' results. Food manufacturers overcome the damage to the cells by freezing very quickly. This is done by exposing the food to the low temperature of liquid nitrogen, to freeze the produce.

— Steam

968.2

Figure 6.23 An experiment to obtain a value for the specific latent heat of vaporisation of water.

Fascinating Physics

Cooling by evaporation

Lindsey went to the swimming pool. After swimming, Lindsey stood at the side of the pool. She felt cold, even though before entering the water she was quite warm.

We have all experienced this effect. When you are dry you feel quite warm, but if you are wet in the same location you quickly feel cold. This is due to the water evaporating off your body. For the water to evaporate, it requires energy which it takes from your skin and so you feel cold.

Refrigerator

The cooling effect of an evaporating liquid is used by a refrigerator. When the cooling liquid (coolant), which boils at a very low temperature (about –30°C), reaches the ice box, it expands rapidly through a small valve and evaporates. To evaporate, the coolant takes in heat energy from the freezer compartment. The freezer compartment, having lost heat becomes cold. The warmed coolant gas enters the compressor which compresses it and pumps it to the back of the refrigerator, where it cools and turns back into a liquid. During the cooling stage, the coolant releases heat energy into the room from the large black fins at the back of the refrigerator which help the coolant to cool to room temperature. The coolant is then allowed to expand and its temperature falls. The coolant is now ready to pass through the icebox to start the cooling cycle again (figure 6.24).

Fascinating Physics

Pressure cooking

At normal atmospheric pressure, water boils at 100°C. However, at pressures higher than atmospheric pressure, water boils at a higher temperature. The steam produced will be at a temperature greater than 100°C. In a pressure cooker, due to the higher pressure, the temperature of the steam is about 115°C. The steam is forced through the food, which cooks more rapidly as a result of the higher temperature. This saves energy.

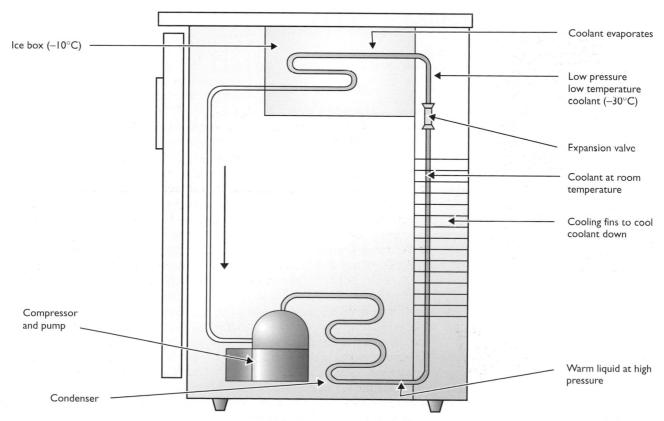

Figure 6.24 A refrigerator.

Section 6.4 Summary

◆ Heat always moves from a hot substance to the cooler surroundings by the processes of conduction, convection and radiation.
◆ Heat loss from a house depends upon the temperature difference between the inside and outside of the house and the time taken.
◆ Equal masses of different substances require different amounts of energy to change their temperature by 1°C.
◆ The energy absorbed or lost by a substance, E_h, is measured in joules (J), specific heat capacity, c, is measured in joules per kilogram per degree Celsius (J/kg°C), mass, m, is measured in kilograms (kg) and the change in temperature, ΔT, is measured in degrees Celsius (°C).
◆ Energy absorbed or lost during a change in temperature = specific heat capacity × mass × change in temperature

$$E_h = c \, m \, \Delta T$$

◆ A specific heat capacity, c, of 100 J/kg°C means that 100 J of energy is required to change the temperature of 1 kg of the substance by 1°C.
◆ A change of state occurs when a solid changes to a liquid (or a liquid changes to a solid) or a liquid changes to a gas (or a gas changes to a liquid).
◆ There is no change in temperature when a change of state occurs.
◆ During a change in state, the energy absorbed or lost by the substance, E_h, is measured in joules (J), mass, m, is measured in kilograms (kg) and the specific latent heat of fusion or vaporisation, l, is measured in joules per kilogram (J/kg).
◆ Energy absorbed or lost during a change in state = mass × specific latent heat

$$E_h = m \, l$$

End of Section Questions

1 Describe **two** ways of reducing heat loss in the home due to:

   (a) conduction; (b) convection; (c) radiation.

2 5000 J of energy are required to raise the temperature of 2 kg of a liquid by 1°C. How much energy will be required to raise the temperature of:

   (a) 4 kg of the liquid by 1°C?
   (b) 6 kg of the liquid by 1°C?
   (c) 4 kg of the liquid by 2°C?
   (d) 8 kg of the liquid by 5°C?

3 What is meant by the statement 'the specific heat capacity of glass is 500 J/kg°C'?

4 How much heat energy is required to raise the temperature of 0.5 kg of water from 20°C to 60°C? The specific heat capacity of water is 4180 J/kg°C.

5 How much heat energy is required to increase the temperature of a 0.40 kg steel pan from 18°C to 88°C? The specific heat capacity of steel is 500 J//kg°C.

6 A kettle with a 2200 W heating element contains 1.2 kg of water. The initial temperature of the water is 15°C. Assuming all the energy is absorbed by the water, find how long it will take for the water to reach 100°C. The specific heat capacity of water is 4180 J/kg°C.

7 A student uses an electrical heater to heat a 1 kg block of aluminium. The current in the heater is 2.0 A and the voltage across the heater is 12 V. The heater is switched on for 12 minutes. The specific heat capacity of aluminium is 902 J/kg°C.

   (a) How much electrical energy was transferred by the heater in this time?
   (b) Calculate the maximum possible rise in the temperature of the aluminium block.

8 How much energy is required to change 2 kg of ice at 0°C into water at the same temperature? The specific latent heat of fusion of ice is 334 000 J/kg.

9 How much energy is required to change 2 kg of steam at 100°C into water at the same temperature? The specific latent heat of vaporisation of water is 2 260 000 J/kg.

10 A 2000 W heater is used to bring 0.8 kg of water to its boiling point of 100°C. The heater is left on for a further 100 s. Calculate the mass of water boiled off in the 100 s. The specific latent heat of vaporisation of water is 2 260 000 J/kg.

Exam Questions

1 The main sources of energy used in the UK are listed below.

Coal, hydroelectric, natural gas, nuclear, oil, wind.

(a) From the above list, select **one** source of energy which is *not* a fossil fuel.

(b) Name a renewable source of energy which is *not* in the list.

(c) Give **one** disadvantage of generating electrical power from the wind.

2 A student investigating the efficiency of a motor sets up the apparatus shown in the figure below.

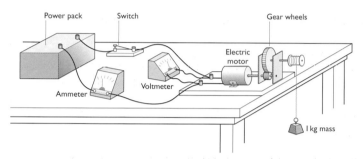

Figure E6.1

When the motor is switched on, a 1.0 kg mass is raised through a vertical height of 0.9 m. The student records the following results during the lifting operation.

voltmeter reading = 6.0 V;
ammeter reading = 4.0 A;
time to lift mass 0.9 m = 2.0 s

(a) Calculate the gravitational potential energy gained by the mass.

(b) Calculate the electrical energy supplied to the motor during the lifting operation.

(c) Find the efficiency of the lifting operation.

(d) Explain why the efficiency of the motor and gears is less than 100%.

3 An anemometer is a device used to measure wind speed. A simple anemometer is shown in the figure below.

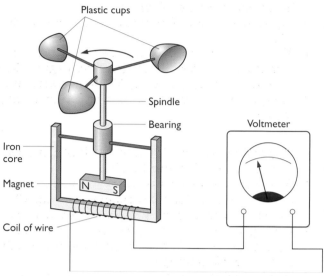

Figure E6.2

When a steady wind blows, the spindle rotates and the pointer on the voltmeter is deflected to give a constant reading.

(a) Is an a.c. voltmeter or a d.c. voltmeter connected to the coil of wire?

(b) When the spindle rotates, a reading is displayed on the voltmeter. Explain why a reading is displayed on the voltmeter.

(c) The wind speed increases. Explain what effect this would have on the reading on the voltmeter.

(d) Give **two** ways in which this anemometer could be modified so as to produce a larger reading on the same voltmeter.

4 The 230 V mains supply is connected to the 1150 turn primary coil of a transformer. A 48 W lamp is connected to the 60 turn secondary coil of the transformer as shown in the figure below.

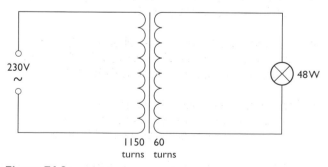

Figure E6.3

The transformer can be assumed to be 100% efficient.

(a) Calculate the voltage across the lamp.
(b) What is in the current in the lamp when it is working at its correct power rating?
(c) Find the current in the primary coil of the transformer.

5 An electric saw is rated at 115 V, 690 W.

(a) Calculate the current in the saw when operating at its stated rating.
(b) The 230 V mains supply is transformed to the 115 V required by the saw using a transformer.
   (i) Is the saw connected to a step-up or step-down transformer?
   (ii) The current in the primary coil of the transformer is 3.1 A. Calculate the efficiency of the transformer.
   (iii) Give one reason why transformers are not 100% efficient.

6 The element of an immersion heater is completely immersed in 1.2 kg of water. The heater is connected to a 12 V supply and switched on. The temperature of the water rises by 12.5°C in 23 minutes. The specific heat capacity of water is 4180 J/kg °C.

(a) Calculate the amount of heat energy gained by the water.
(b) Estimate the power rating of the heater, stating any assumption you make. (Use an appropriate number of figures in your answer.)
(c) Find the current in the element of the heater.

7 A sample of a substance is at a temperature of 20°C. The sample is in the solid state. The mass of the sample is 0.4 kg. A heater, rated at 100 W, is used to heat the solid sample for 1100 s. The graph shows how the temperature of the substance varies with time.

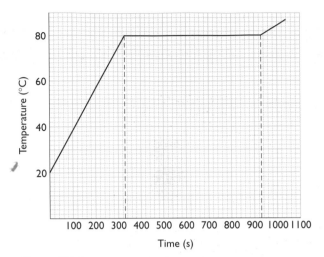

Figure E6.4

(a) What is the melting point of the substance?
(b) Calculate the specific heat capacity of the substance in the solid state.
(c) Calculate the specific latent heat of fusion of the substance.

CHAPTER SEVEN
Space Physics

Figure 7.1

Figure 7.2 After an explosion in the service module, the crew of Apollo 13 were pushed safely back to Earth by the Apollo 13 lunar module.

'To boldly go where no one has gone before'.

Opening credit for Star Trek

'I believe this nation should commit itself to achieving the goal, before this decade is out, of landing a man on the Moon and returning him safely to Earth. No single space project in this period will be more impressive to mankind, or more important for the long-term exploration of space; and none will be so difficult or expensive to accomplish'.

President John F Kennedy, 25 May 1961

Neil Armstrong, Commander Apollo 11 Lunar Lander, 1969 (figure 7.1): 'Houston, Tranquility Base here, The Eagle has landed'.

Mission Control: 'Roger, Tranquility, we copy you on the ground. We're breathing again. Thanks a lot'.

Jim Lovell, Commander Apollo 13 (figure 7.2): 'Houston, we have a problem'.

Since the beginning of time people have dreamed of going into space and of travelling to different planets. Since 1969 it has been possible to reach the moon and it is possible that one day people will travel to Mars. But we can observe signals from space without travelling there since we can use telescopes and other instruments to make both measurements and observations.

At the end of this section you should be able to:

1 Use correctly in context the following terms: moon, planet, sun, star, solar system, galaxy, universe.
2 State approximate values for the distance from the Earth to the Sun, to the next nearest star, and to the edge of our galaxy in terms of the time for light to cover these distances.
3 Draw a diagram showing the main features of a refracting telescope: objective, eyepiece, light-tight tube.
4 State that the objective lens produces an image which is magnified by the eyepiece.
5 State that different colours of light correspond to different wavelengths.
6 List the following colours in order of wavelength: red, green, blue.
7 State that white light can be split into different colours using a prism.
8 State that the line spectrum produced by a source provides information about the atoms within the source.
9 State that there exists a large family of waves with a wide range of wavelengths which all travel at the speed of light.
10 State that telescopes can be designed to detect radio waves.

Seeing afar

Terms used to talk about space have a special meaning:

◆ A **planet** is an object that orbits a star. It reflects light and produces no energy.
◆ A **moon** is an object that orbits a planet. It reflects light.
◆ A **star** is a ball of gases which produces heat and light.
◆ A **galaxy** is a system of stars and dust that is both spinning and travelling (figure 7.3).
◆ The **Universe** is the whole of space which can be detected.

We live on a planet called Earth. The Earth is the third of nine planets which orbit around the Sun. The Sun is a star which glows giving off both heat and light energies. The nine planets circle around the Sun and form the Solar System (figure 7.4). The Solar System is a very small part of thousands of millions of stars which form part of a galaxy called the Milky Way. In turn there are millions of more galaxies which together form the Universe.

Figure 7.3 Andromeday: one of the many galaxies in the Universe.

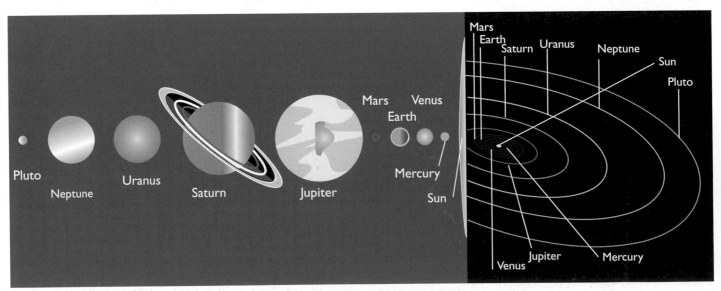

Figure 7.4 Our Solar System.

The light year

The light year is the distance travelled by a beam of light in one year. Light years are required as our unit for measuring distance as the distances involved in space are so large.

$$\text{Speed of light in space} = 3 \times 10^8\,\text{m/s}$$
$$\text{Distance travelled by light in 1 second} = 3 \times 10^8\,\text{m}$$
$$\text{Number of seconds in one hour} = 3600$$
$$\text{Number of seconds in one day} = 3600 \times 24$$
$$\text{Number of seconds in one year} = 3600 \times 24 \times 365$$
$$\text{Distance travelled by light in one year} = 3600 \times 24 \times 365 \times 3 \times 10^8\,\text{m}$$
$$= 9.5 \times 10^{15}\,\text{m}$$
$$\text{One light year} = 9.5 \times 10^{15}\,\text{m}$$

This distance is far too large for human beings to understand so you can see that it is far easier to use the term light year to help us understand the large distances that are involved.

Table 7.1 shows the time taken for light to travel to us from distant objects, and the distance and the light years it covers in that time.

Source	Time taken for light to reach Earth	Distance travelled	Number of light years
Moon	1.2 seconds	$3.6 \times 10^8\,\text{m}$	3.8×10^{-8}
Sun	8 minutes	$1.4 \times 10^{11}\,\text{m}$	1.5×10^{-5}
Nearest star (after the Sun): Proxima Centauri	4.3 years	$4.1 \times 10^{16}\,\text{m}$	4.3
Other side of our galaxy	100 000 years	$9.5 \times 10^{20}\,\text{m}$	100 000
Andromeda galaxy (our nearest galaxy)	2 200 000 years	$2.1 \times 10^{22}\,\text{m}$	2 200 000

Table 7.1 The Earth's relationship to various points in space

Light and its components

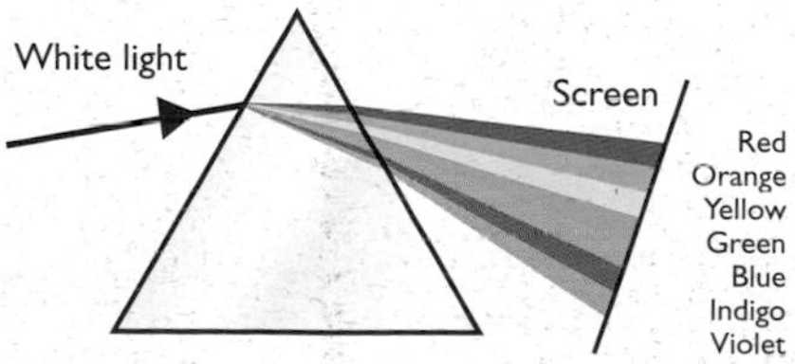

Figure 7.5 White light passing through a prism.

In Chapter 3, It was shown how the different parts of the electromagnetic spectrum were used in health physics. The use of light from lasers was mentioned. It is possible to show that white light consists of a number of different colours. This can be done by passing white light through a glass triangular prism. The white light splits into a **spectrum** of colours. Each colour has a different frequency and wavelength. At one end of the spectrum, there is red. It has the longest wavelength. This is shown in figure 7.5. The colours after red are orange, yellow, green and blue. At the other end there is violet which has the shortest wavelength. Blue light is refracted more than red. All the colours travel at the same speed which is the speed of light.

Electromagnetic spectrum

In space, light is not the only wave that travels at a speed of $3 \times 10^8\,\text{m/s}$. Figure 7.6 lists waves which also travel at $3 \times 10^8\,\text{m/s}$ (300 000 000 m/s). You met this electromagnetic spectrum of waves in Chapter 3. Each radiation has a different wavelength and will need a specific detector. Each radiation is shown in the figure with a suitable detector.

	Short wavelength						Long wavelength
Radiation	Gamma rays	X-rays	Ultra-violet	Visible light	Infra-red	Micro-waves	TV and radio
Detector	Photo film	Photo film	Fluorescent material	Eye	Photo transistor	Aerial	Aerial
	High frequency						Low frequency

Figure 7.6

All the information and facts that we know about space have been discovered by detecting signals given out by stars throughout the Universe. These signals have different wavelengths (and frequencies) and so are different members of the electromagnetic spectrum (since they all have a speed of 3×10^8 m/s in air). To pick up some of these different types of signals, different kinds of telescopes are used.

Detecting the signals from space

Radio telescopes

Large unpolished metal dishes, often formed of mesh wire, collect and direct the weak radio waves onto an aerial at the focus of the dish. The aerial feeds the signals to an amplifier. Radio waves have a large wavelength (0.001 m to 1000 m) and so the collecting surface need not be accurately shaped (figure 7.7).

To see fine detail, the opening of a telescope should be as large as possible. Large enough openings cannot be achieved in a single-dish radio telescope. Several small dishes in a line are used. The results are computerised, and openings of up to 5 km diameter can be simulated (figure 7.8).

Figure 7.7 Radio telescopes.

Microwaves

Astronomers can detect radiation from space which travels at the speed of light and has a wavelength of several millimetres. This is called microwave radiation. The wavelength used in a microwave oven at home is about 12 cm. This radiation can give them information about the temperature of stars and this tells astronomers about the age of the stars since, as the star reactions change, its temperature will change as time goes on.

Infrared radiation

Infrared radiation arrives at the Earth from objects in space and provides astronomers with another source of information. Infrared radiation has a longer wavelength (lower frequency) than visible light and shows up as heat. This is discussed later in this chapter.

Figure 7.8 A row of dish antennae which make up the Very Large Array in New Mexico, USA, the World's largest radio telescope array.

Optical refracting telescopes (light telescopes)

Stars are so far from the Earth we need to have telescopes to view them. A basic telescope consists of a long tube with a lens at either end. The lens you look through is the eyepiece and the other lens is called the objective (figure 7.9).

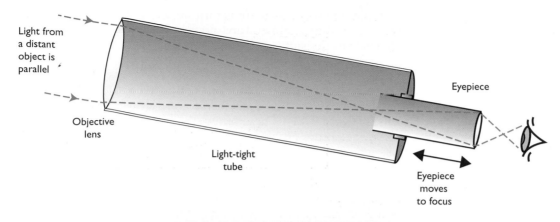

Figure 7.9 Lenses in a telescope.

- The objective lens has a long focal length and the light from a distant star is brought to a focus part way down the tube.
- This image is then magnified by the eyepiece lens. The final image is upside down.
- The eyepiece acts as a magnifying glass (figure 7.10). The rays coming from the object or star travel in straight lines. To help with the diagrams we only need to consider two lines. One ray travels straight through the centre of the lens. The other passes parallel to the axis of the lens and then straight through the focal point. The rays do not meet unless they are traced back.

This image from the eyepiece is called a virtual image since it cannot be displayed on a screen.

Magnification and image brightness

The magnification of an image can be found by measuring the height of the image from the eyepiece lens and dividing by the height of image from the objective lens.

$$\text{Magnification} = \frac{\text{height of image from eyepiece lens}}{\text{height of image from objective lens}}$$

If the magnification of the final image is 3, the height of the image is three times that of the object. How clear the final picture of the star is also depends on its brightness.

The brightness of the star decreases as its diameter decreases. The broader the star is, the brighter it is and so the easier it is to view it with an optical telescope.

To obtain as bright an image as possible the diameter of the objective lens should be as large as possible so that as much light as possible can be collected.

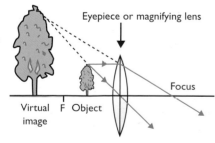

Figure 7.10

Other telescopes will detect infrared radiation as heat. This provides an indication of the temperature of stars which can be used to tell us about the life cycle of the star.

The Hubble images

To detect light from distant objects it is necessary to detect as much light as possible. In most telescopes this is done by using curved mirrors instead of lenses (figure 7.11). These reflect the light from the mirror to a focus. This was done on the Hubble telescope but the original curvature of the mirror was incorrect and a space mission was required to insert a lens to correct the fuzziness of the image (figure 7.12). Telescopes such as Hubble are used in space to overcome the problems of looking through the Earth's atmosphere. Light pollution from nearby towns and the general air pollution produce poorer images. But new computer techniques will correct these problems and the great cost of space telescopes has meant that fewer are now being launched. A selection of images from Hubble is shown in figure 7.13.

Fascinating Physics

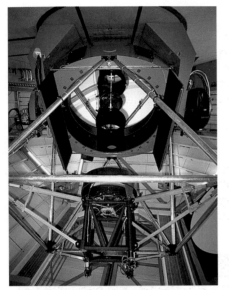

Figure 7.11 A curved mirror telescope.

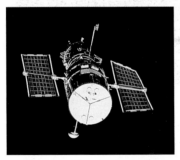

Figure 7.12 Hubble telescope.

Figure 7.13 Images from Hubble telescope.

Temperature of stars

Colour and temperature

When substances are strongly heated, a whole range of wavelengths are usually given out to form a continuous spectrum. The amount of light of each colour depends on the temperature of the substance. If you heat a piece of metal in a very hot flame then it changes from red hot to white heat to blue heat, that is a change of temperature is shown as a change of colour. As the temperature increases, the wavelength of light emitted becomes smaller.

Astronomers can judge the temperature of the surface of stars by noting the colour of light which the stars give out (figure 7.14). The colour of a star depends on its surface temperature.

Some familiar stars can be placed in order of increasing temperature: Barnard's Star – red; Betelgeuse – orange-red; the Sun – yellow; Rigel – bluish-white. Look at the galaxy called M15 photographed by the Hubble space telescope (figure 7.15). You should be able to determine the star which is blue heat – this star will be the one at the highest temperature.

Figure 7.14 Stars give out different amounts of colour in the light they emit, depending on the star's surface temperature.

Figure 7.15 Galaxy MIS.

Type of stars

Astronomers can identify the elements that are present in stars by carefully examining the spectrum produced when the light from a star is viewed through a prism. The spectrum from a star is a line spectrum made up from the spectra (colours) of all the elements present.

Line spectra

When an electric current passes through a gas, it gives energy to the gas. This energy is then given out as light of several definite wavelengths (colours). This is called a **line emission spectrum**. Each element has its own particular spectrum by which it may be identified. This enables astronomers to identify elements present in distant stars. For example, a spectrum which identified sodium would have two distinct yellow lines, which are unique to sodium.

Spectrum

You can identify elements which are present in several simplified stellar spectra as follows. The lines from some of the known elements should match against some of the lines from the star's spectrum. You should repeat this for each known element until you have identified all the elements that make up this star (figure 7.16).

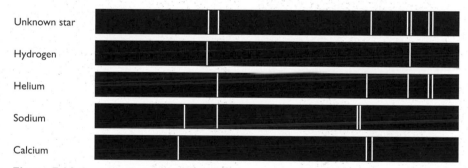

Figure 7.16 Spectra of an unknown star.

By examining this spectrum we can say that the star in figure 7.16 contains hydrogen and helium since the lines in the spectra coincide. Figure 7.17 shows the spectra of a star.

Figure 7.17 Spectra of a star.

End of Section Questions

1 Draw a diagram showing the key features of an astronomical telescope. In your diagram label the two lenses.

2 The mariner missions were sent to explore Mars.

   Suggest a reason why this same probe could not be used to explore Jupiter, where the atmospheric conditions are not similar since the surface temperature is much greater.

3 The diagram below shows a converging lens which is used as an eyepiece of a telescope.

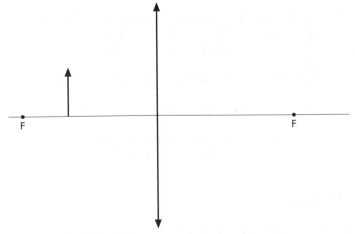

Figure 7.1Q3

(a) Copy and complete the diagram to show the final image produced.
(b) State **two** differences between the final image and the object.

4 Name a detector for: (a) ultraviolet; (b) infrared radiation.

5 Explain the purpose of the large curved part of a radio telescope.

6 Spectra from stars are used by astronomers to give information about the star.

   (a) Describe the kind of information about the star obtained from such spectra.
   (b) A star X15 has a red-orange appearance but the star Y17 has a blue white appearance. What can you tell from this information about the two different stars?

7 (a) The light from Alpha Centauri takes 4.3 light years to reach Earth. If this light was emitted on 20 July 1969 (the day of the lunar landing) what was the month and the year that it was seen on Earth?
   (b) State the distance in light years to the edge of our galaxy.

Jets and rockets

At the end of this section you should be able to:

1 State that a rocket is pushed forward because the 'propellant' is pushed back.
2 Explain simple situations involving the rule: A pushes B, B pushes A back.
3 Carry out calculations involving thrust, mass, and acceleration.
4 Explain why a rocket motor need not be kept on during interplanetary flight.
5 State that the force of gravity near the Earth's surface gives all objects the same acceleration (if the effects of air resistance are negligible).
6 State that the weight of an object on the moon or on different planets is different from its weight on Earth.
7 State that objects in free fall appear weightless.
8 Explain the curved path of a projectile in terms of force of gravity.
9 State that an effect of friction is the transformation of E_k into heat.

10 State that Newton's Third Law is: 'If A exerts a force on B, B exerts an equal but opposite force on A'.
11 Identify 'Newton pairs' in situations involving several forces.
12 Explain the equivalence of acceleration due to gravity and the gravitational field strength.
13 Carry out calculations involving the relationship between weight, mass, acceleration due to gravity and/or gravitational field strength including situations where g is not equal to 10 N/kg.
14 Use correctly in context the following terms: mass, weight, inertia, gravitational field strength, acceleration due to gravity.
15 State that the weight of a body decreases as its distance from the Earth increases.
16 Explain how projectile motion can be treated as two independent motions and solve numerical problems using this method.
17 Explain satellite motion as an extension of projectile motion.
18 Carry out calculations involving the relationships $E_h = cm\Delta T$, work $= Fd$ and $E_k = \frac{1}{2}mv^2$.

Jets

Jet engines in aircraft work using physics that we use in our everyday life. The following are examples of the ideas behind their operation.

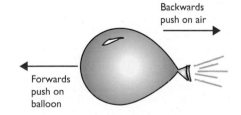

Figure 7.18

◆ If you push against a wall, the wall pushes you backwards.
◆ To stand up, you push down.
◆ In walking, you push your foot backwards but the floor pushes you forwards.
◆ In swimming, you push the water backwards but you are pushed forward.
◆ If you release a blown up balloon with its neck open, the air is pushed out the back but the balloon is pushed forward (figure 7.18).

Newton's Third Law

These are all examples of Newton's Third Law of Motion (Newton III) which states that

To every action, there is an equal but opposite reaction.

Put simply, this means that when A pushes B, B pushes A back with the same size of force. In the example above, the balloon pushes the air to the left, so the air pushes the balloon to the right.

> Action: *balloon* pushes *air* to *left*.
> Reaction: *air* pushes *balloon* to *right*.

Newton's Third Law is the principle behind how rockets and jet engines are able to produce motion. In both cases a high-speed stream of hot gases (produced by burning fuel) is pushed backwards from the vehicle with a large force (figure 7.19) and a force of the same size pushes the rocket forwards.

> Action: *vehicle* pushes *hot gases backwards* (downwards).
> Reaction: *hot gases* push *vehicle forwards* (upwards).

For the fuel to burn and produce hot gas (exhaust gas) a supply of oxygen is needed. Space rockets carry their own oxygen supply but jet engines use the oxygen in the surrounding air. This means that rockets can fly in space (a vacuum so no oxygen) but jet aircraft cannot.

Rockets

In 1926, Robert Goddard, an American physicist, developed a liquid propellant rocket which burned petrol and liquid oxygen. He developed a great deal of the basic technology of rocket motion. During the Second World War, a great deal of rocket development took place, mainly by the scientist Werner von Braun. After the war, the development of rockets concentrated on putting objects into space, mainly satellites and different telescopes, but also people. For a rocket to escape the Earth's gravitational pull, it must reach a speed of more than 11 000 m/s. Modern rockets use liquid hydrogen as the fuel. To support the tremendous rate of fuel-burn required, a large supply of oxygen is needed. This is carried in the form of liquid oxygen.

A simple rocket is shown in figure 7.20. The two liquids mix and burn in the combustion chamber, where the hot gases produced expand rapidly and are forced through the nozzle. These hot gases are pushed downwards and exert an upwards force on the rocket (from Newton's Third Law). As long as the force is greater than the weight of the rocket, lift off can take place.

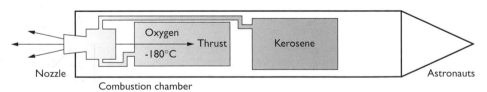

Figure 7.20 A simple design for a rocket. The rocket carries with it fuel, such as kerosene, and liquid oxygen. These are mixed and burnt in the combustion chamber.

Apollo missions

The era of human missions to the Moon is remarkably short. The Apollo missions started in February 1967 and Apollo 11 landed in July 1969. The last mission was Apollo 17 which took place in December 1972. The Apollo missions were launched by Saturn V rockets which were over 100 m high and had a mass of more than 3 million kg (figure 7.21). The first stage burned a mixture of kerosene and liquid oxygen for 160 seconds. The second stage burned liquid hydrogen and liquid oxygen for 6.5 minutes. The last stage burned for 160 seconds and put the vehicle into a circular parking orbit.

The missions allowed 12 men to walk on the surface of the Moon and over 400 kg of moon rock were collected. The plaque left on the Moon on the last mission read:

Figure 7.19 An Atlas rocket. The burnt fuel is driven away from the rocket with great force. In turn, the fuel gives an equal and opposite force to the rocket, which drives it into space.

Figure 7.21 Saturn V rocket.

Fascinating Physics

'Here man completed his first exploration of the Moon December 1972. May the spirit of peace in which we came be reflected in the lives of all mankind.'

Ditching

When a space rocket is launched, the force that the exhaust gas exerts on the rocket causes the rocket to accelerate upwards. The Earth's gravitational pull will prevent the rocket from escaping from the Earth, unless the rocket is travelling fast enough. To reduce the amount of fuel required, a clever trick is used. When a rocket reaches a certain speed, its mass is reduced. This is usually done by releasing empty fuel tanks or the lower stage of the rocket. The rocket motor now exerts a force on a smaller mass and therefore its acceleration will increase.

Taking Off

Close to the Earth's surface the gravitational pull (force of gravity or weight of the rocket) on the rocket is very large. This means that a large force (or thrust) must be exerted by the rocket engines (figure 7.22).

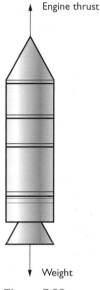

Engine thrust

Weight

Figure 7.22

$$\text{Unbalanced force} = \text{engine thrust} - \text{weight of rocket}$$

$$\text{Acceleration of rocket} = \frac{\text{unbalanced force}}{\text{mass of rocket}}$$

The Saturn V rocket which launched the Apollo missions to the Moon had a total mass at lift-off of 3×10^6 kg. The rocket motors could exert a thrust of 3.3×10^7 N.

$$\text{Weight of rocket} = mg$$
$$= 3\,000\,000 \times 10$$
$$= 30\,000\,000\,\text{N}$$

$$\text{Unbalanced force } F_{\text{un}} = \text{engine thrust} - \text{weight of rocket}$$
$$= 33\,000\,000 - 30\,000\,000$$
$$= 3\,000\,000\,\text{N}$$

$$a = \frac{F_{\text{un}}}{m} = \frac{3\,000\,000}{3\,000\,000} = 1\,\text{m/s}^2$$

In outer space, where the gravitational field strength is zero, the weight of the rocket is zero. In this case:

$$\text{Unbalanced force on rocket} = \text{engine thrust}$$

You should note that the effect did not occur with the Apollo spacecraft since there was never a chance that it could travel to deep space.

When fired to reposition the shuttle in space, the rocket engine can produce a thrust of 10 000 N. The shuttle has a total mass of 75 000 kg.

$$\text{acceleration of shuttle} = \frac{\text{unbalanced force}}{\text{mass}}$$
$$= \frac{\text{engine thrust}}{\text{mass}}$$
$$= \frac{10\,000}{75\,000}$$
$$= 0.13\,\text{m/s}^2$$

Fascinating Physics

Rocket technologies for the future

Rockets using liquid fuels are at present the only method humans have to produce the large forces required. But if we are to travel to other planets then other forms of transport must be developed. The latest work examines ion engines. Ions were described in Chapter 3. This, it is hoped, will allow much smaller engines which can develop large amounts of thrust. Another development looks at the ideas of magnetic particles which may produce similar forces. If humans are to travel further into space then different technologies must be explored.

The Space Shuttle

The key facts about the Shuttle are shown in figure 7.23. The Space Shuttle is a re-usable craft that can carry payloads and crews to and from space. On the launch pad it is joined to two solid fuel rocket boosters and a large external fuel tank. The boosters help lift the craft away from the Earth's gravitational pull. The boosters then fall away into the ocean where they can be retrieved for further use. The external tank is not re-used and holds a total of over 700 000 kg of fuel. The fuel from this tank is used to position the shuttle to a point just short of its final orbit. The tank then separates and falls back into the ocean. The Shuttle coasts for a few seconds and then fires its two orbital engines to put it into the correct position for its orbit. At launch you would hear the immense roar of the engines. You would experience an acceleration of about three times that of the pull of gravity. (This is about the same level of discomfort as if you turn a corner much too fast in a car.) This happens for a short time near the two minute mark just before the two solid fuel boosters burn out and drop off.

About five minutes later the liquid tank empties and separates and re-enters the atmosphere. This happens about 185 km from the Earth (figure 7.24).

The Space Shuttle or the space transportation system (STS), is launched vertically. At launch, it has a total mass of 2.05×10^6 kg and the thrust developed by the engines is 2.1×10^7 N. Once in orbit, the cargo doors open to allow experiments to take place, satellites to be launched and observations to be made (figure 7.25). The duration of flight depends on the nature of the experiments to be carried out. In orbit, the orbital manoeuvering system

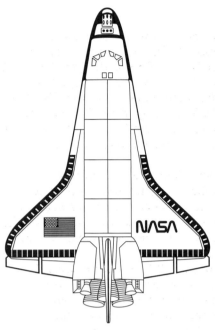

Figure 7.23

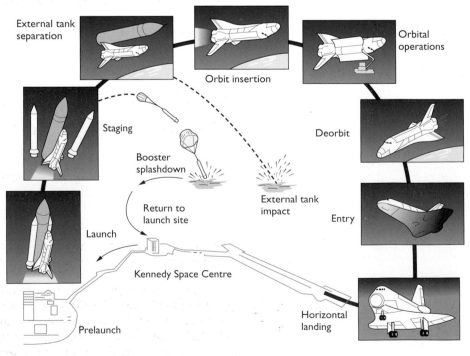

Figure 7.24 Space Shuttle typical mission profile.

(OMS) engines can be used to alter the orbit of the shuttle orbiter. When all experiments are complete, the cargo doors are closed and the crew prepare for re-entry into the Earth's atmosphere. The Shuttle is about the same length and weight as a commercial aircraft but its appearance is very different.

Re-entry

Before re-entry each astronaut puts on a special suit which is really an inflatable pair of pants. This prevents the blood from pooling into the lower part of the body and causing blackout. One hour before landing the speed of the Shuttle is reduced by turning it backwards. The two small engines ignite and the burn takes it out of its orbit and the Shuttle begins to drop. While it falls the pilot turns it round again and pulls its nose up. After 30 minutes the Shuttle has entered the Earth's atmosphere at a height of 120 km. It will then be programmed by the computers to cover a distance of 6400 km and drop 120 km in the next 30 minutes.

When the shuttle re-enters the atmosphere there is a communications blackout. As it re-enters the Earth's atmosphere the intense heat produced due to the work done by friction will cause the temperature to reach about 1650°C on the surface of the nose and the leading edges of the wings. As the craft slows down it will reach a height of 23 km and be travelling at 550 m/s. The Shuttle will be lined up with the flight path.

The approach will start at 4000 m and there will be no noise since the shuttle will land like a glider. Its speed will now be about 150 m/s and its angle of descent will be about 7 times that of an ordinary aircraft. At a height of 500 m it levels off. The landing wheels come down at a height of 90 m and the speed on touch down will be about 90 m/s (figure 7.27).

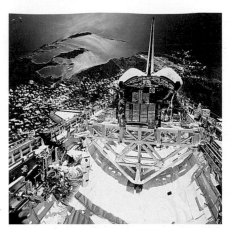

Figure 7.25 Orbital operations on the Shuttle.

Figure 7.26 The Shuttle landing.

Acceleration due to gravity

An object falling near the surface of the Earth accelerates (if the effects of air resistance are negligible). This acceleration is called the acceleration due to gravity (g).

Consider two objects of mass 2 kg and 7 kg falling near the surface of the Earth. We are assuming that there is no friction present.

$$
\begin{aligned}
\text{Weight} &= \text{force of gravity} \\
&= mg \\
&= 2 \times 10 \\
&= 20 \text{ N}
\end{aligned}
\qquad
\begin{aligned}
\text{Weight} &= \text{force of gravity} \\
&= mg \\
&= 7 \times 10 \\
&= 70 \text{ N}
\end{aligned}
$$

Since we are assuming no friction is present:

$$
\begin{aligned}
\text{Unbalanced force} = F_{un} &= \text{weight} \\
&= 20 \text{ N} \\
a = \frac{F_{un}}{m} &= \frac{20}{2} \\
&= 10 \text{ m/s}^2
\end{aligned}
\qquad
\begin{aligned}
\text{Unbalanced force} = F_{un} &= \text{weight} \\
&= 70 \text{ N} \\
a = \frac{F_{un}}{m} &= \frac{70}{7} \\
&= 10 \text{ m/s}^2
\end{aligned}
$$

This shows clearly that the acceleration due to gravity in the absence of friction (air resistance) is the same for *all* objects no matter what their mass is.

The acceleration due to gravity on a planet has the same number value as the gravitational field strength on the planet. For example,
◆ On a planet: acceleration due to gravity = gravitational field strength
◆ On Jupiter: gravitational field strength = 25 N/kg
    acceleration due to gravity = 25 m/s²

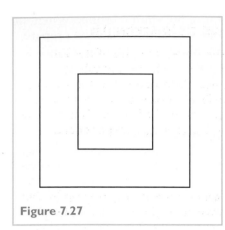

Figure 7.27

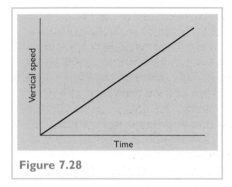

Figure 7.28

Measuring the acceleration due to gravity

Using a light gate, computer and a piece of card cut as a mask, we can calculate the acceleration due to gravity. The card is dropped through the light gate, from different heights, and the acceleration measured in each case (figure 7.27).

The height the object is released from does not affect the acceleration. For the Earth, the acceleration due to gravity, $g = 10\,\text{m/s}^2$

The speed–time graph for an object dropped vertically, from rest, near the Earth's surface (ignoring air resistance) is shown in figure 7.28.

Weight on Earth and other planets

The weight of an object near the surface of a planet such as the Earth is the pull of the planet on the object. Like all forces, the weight is measured in newtons.

The pull of gravity on a falling object (the weight) can be calculated using the equation:

$$W = mg$$
weight = mass × gravitational field strength

and the value for g on Earth = $10\,\text{N/kg}$.

The pull of gravity per kilogram mass on a planet is called its gravitational field strength. Gravitational field strength is measured in newtons per kilogramme (N/kg).

Table 7.2 shows the weight of different objects on the Earth. Table 7.3 shows the weight of a 1 kg mass on different planets and the Moon.

Weight (N)	Mass (kg)
10	1
300	30
4	0.4
1	0.1
500	50
0.3	0.03

Table 7.2 Weight of objects on Earth

Planet	Weight of 1 kg (N)	Gravitational field strength (N/kg)
Mercury	4	4
Venus	9	9
Earth	10	10
Mars	4	4
Jupiter	25	25
Saturn	10	10
Uranus	10	10
Neptune	12	12
Moon	1.6	1.6

Table 7.3 Weight of 1kg on different planets and the Moon

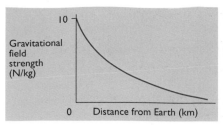

Figure 7.29 Change in gravitational field strength.

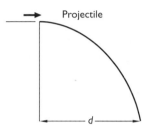

Figure 7.30 Projectile motion.

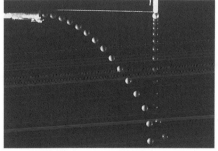

Figure 7.31

Effect of distance on gravitational field strength

As a spacecraft rises from the surface of the Earth, the pull of the Earth's gravity on it gets smaller and smaller. Figure 7.29 shows a graph of gravitational field strength, g against distance. Astronauts in orbit around Earth (usually about 200 km above the surface) appear to be weightless. True weightlessness can happen only where the pull of gravity is zero. This can only happen in deep space – which no humans have experienced.

Projectile motion

There is a wide variety of objects in orbit around the Earth, ranging in size from satellites the size of a house, to an astronaut's glove. To understand how it is possible for these objects to stay in orbit, we study objects moving horizontally and vertically at the same time. These are called projectiles.

If an object is projected horizontally, at 16 m/s then its inertia (unwillingness to change its motion) should tend to keep it moving at 16 m/s horizontally (in a straight line). Due to the force of gravity (the object's weight) this is not possible and the object experiences a force pulling it downwards while it tries to move horizontally at constant speed. To overcome this conflict between its inertia and its weight, the object follows a curved path (figure 7.30). This is called **projectile motion**. The motion of the projectile is therefore made up of two separate motions:
1 A vertical motion under the influence of the force of gravity.
2 A horizontal motion under the influence of its inertia.
A special photograph is taken of a ball being projected, and a ball being dropped vertically at the same time, as shown in figure 7.31.

The vertical motion of the projected ball and the free falling ball are in step all the way down. This means that the vertical motion of the projectile is the same as a free falling body, that is both are accelerating with a downward acceleration of $10 \, \text{m/s}^2$. This the acceleration due to gravity which was discussed earlier.

The horizontal spacing between the images of the projectile are all equal. This means that the horizontal motion of the projectile is at a constant speed and is equal to the speed of projection.

The motion of a projectile can be treated as two independent motions:
1 Constant speed in the horizontal direction.
2 Constant acceleration in the vertical direction due to the force of gravity.
These two motions are shown in figure 7.32.

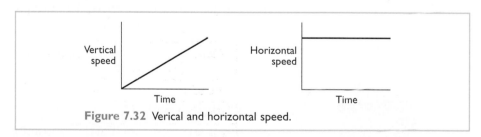

Figure 7.32 Verical and horizontal speed.

When solving problems, we can treat the motion in the horizontal direction *separately* from the motion in the vertical direction.

Example
A flare is fired horizontally out to sea from a cliff top. The speed of the flare is 40 m/s. The flare takes 4 s to reach the sea.
(a) What is its horizontal speed after 4 s ?
(b) Calculate its vertical speed after 4 s.
(c) (i) Draw a graph of vertical speed against time
    (ii) Use the graph to calculate the height of the cliff top.

Solution
(a) Horizontal speed remains at 40 m/s.
(b) Using $v - u = at$ for the vertical motion where $u = 0$, $a = g = 10 \text{ m/s}^2$, and $t = 4$ s:

$$v - 0 = 4 \times 10$$

$$= 40 \text{ m/s}$$

(c) (i) The graph is shown in figure 7. 33.
   (ii) The height of the cliff is the area under the graph.

$$\text{Distance} = \frac{1}{2} \times \text{base} \times \text{height}$$

$$= 0.5 \times 4 \times 40$$

$$= 80 \text{ m}$$

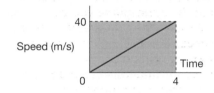

Figure 7.33

Newton's satellite

Newton, in considering the motion of the Moon around the Earth carried out the following 'though experiment'. Suppose a bullet is fired horizontally from a gun situated on top of a high mountain. The bullet will have two motions which occur simultaneously:
1 A horizontal motion at uniform speed (if air resistance is negligible).
2 A vertical motion of a uniform acceleration 'g' vertically downwards under the action of the gravitational attraction between the bullet and the Earth.

As a result, the bullet will follow a curved path. As long as the Earth is flat, the bullet will hit the ground. This will not depend on the horizontal speed of the bullet. However, the approximation of the 'flat' Earth is only valid over a limited range. For a 'round' Earth it becomes important to take the curvature of the Earth into account for projectiles of long range (figure 7.34).

If there was no force of gravity, the bullet would follow the path AB because if no forces act on it, it must travel with a uniform speed in a straight line. Due to the gravitational field of the Earth, the bullet falls continuously below this line. If the bullet falls below the line AB faster than the Earth's surface curves away under it, the bullet will still hit the Earth, e.g. at the point C or D according to the bullet's speed.

If the bullet were fired at such a speed that it fell vertically at the same rate as the Earth's surface curved away under it, then it would always be at the same height above the surface. The bullet would then never reach the surface, but would circle it at a constant altitude. This was the first theory of the artificial satellite.

Remember Newton 'thought ' about this experiment long before we had rockets or satellites.

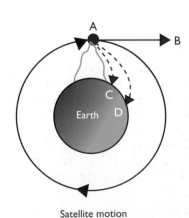

Satellite motion

Figure 7.34 Newton's thought experiment. When an object is projected at the correct speed it never reaches the ground because the Earth is curved, so it remains in orbit free-fall.

Weightlessness

When astronauts are floating around a space vehicle we say that they are weightless (figure 7.35).

Weight is a force which pulls us down towards the ground. Using Newton's Third Law, the ground we are standing on exerts an equal but opposite force on us and we feel the effect of it supporting us. We are made aware of our weight because the ground (or whatever supports us) exerts an upward push on us as a result of the downward push our feet exert on the ground. This is our apparent weight.

When you stand in a lift, you push down on the floor. Using our knowledge of forces, the floor pushes you back up. If the lift is at rest or moving up or down at constant speed there are no extra pushes or pulls on you, so your weight appears to remain unchanged. If the lift accelerates

Figure 7.35 Weightlessness.

upwards, the floor will push you up with a greater force. The reaction is that you will push down with a larger force and your apparent weight will increase. Your real weight is unchanged. If the lift accelerates downwards, the floor will push you up with a smaller force. Again you will push down with a smaller force and your apparent weight will decrease.

If our feet were unsupported we would experience 'weightlessness'. Passengers in a lift with a continual downward acceleration of $10\,\text{m/s}^2$ would get no support from the floor, since they, too, would be falling with the same acceleration. There is no upward push on them and so no sensation of weight is felt. Our apparent weight is zero but our real weight is unchanged. This is experienced briefly when we jump off a wall or dive into a swimming pool.

If an astronaut is in space at a height of a thousand kilometres above the surface of the Earth, the Earth's gravitational field strength (about $7.3\,\text{N/kg}$) is still strong enough to exert a force on him and pull him towards the floor of the spacecraft. However, at the same time, there is a force acting on the spacecraft accelerating it downwards, so the floor of the spacecraft tends to fall away from the astronaut. If the floor of the spacecraft falls away from the astronaut *at the same rate as he is falling*, the floor will not be in contact with the astronaut. The floor exerts no force on the astronaut so that he feels as if he is weightless.

The astronaut and the spacecraft are two projectiles in a continual state of free fall, *both accelerating towards the Earth with the same acceleration*, this state is called **weightlessness**.

True weightlessness can only be experienced by a spacecraft in an area where the gravitational field strength is equal to zero. This is deep in space.

Re-entry and friction

When a spacecraft re-enters the Earth's atmosphere, it is travelling at about $11\,100\,\text{m/s}$. It collides with the air particles of the atmosphere causing a large force of friction on the spacecraft, which is slowed down. The work done by friction reduces the kinetic energy of the spacecraft. This kinetic energy is changed into heat energy. Some of the heat energy is 'lost' to the surroundings. However, some of the heat energy produced is absorbed by the spacecraft and is enough to raise the temperature to around 1300°C.

The space Shuttle makes an unpowered glide through the atmosphere and lands on a runway using its wheels just like an ordinary aircraft. Since it has no power it can only attempt this in one try. Other returning spacecraft use parachutes to slow the spacecraft down and make a gentle 'splashdown' in the sea or on the land

Spacecraft design

Apollo

The heat shield on the Apollo craft was made from a honeycomb of stainless steel to which an epoxy resin coating was bonded. Apollo hit the atmosphere at an angle and this allowed the thickness of the heat shield to be varied to cut down on the overall weight of the craft. Pitch control rockets rotated the craft to ensure that the thickest part of the shield hit the atmosphere first as the intense heat could have burned through the thinner sections of the shield. Some lift was also obtained from the shape of the capsule to enable the re-entry corridor to be selected. Vaporising metal from the shield carried away most of the heat reducing its thickness to less than a centimetre (figure 7.36).

Shuttle

The Shuttle is made from aluminium alloy covered in special tiles to protect it from the intense heat. Each Shuttle orbiter will require almost 34 000 thermal protection tiles in different sizes, shapes and thicknesses. The material of the tiles must be able to withstand repeated heating and cooling plus extreme noise and vibrations for up to 100 flights without replacement.

Figure 7.36 Apollo 8 heat shield.

The material is a high purity silica compound that sheds heat so quickly that a piece of material can be held in an ungloved hand only seconds after removal from an extremely hot oven while the interior of the tile still glows red hot! Tiles for the underside of the Shuttle and other areas exposed to high temperatures up to 1300°C receive a black glass coating. For areas exposed to lower temperatures (up to 700°C) a white silica compound is used with shiny alumina oxide added to reflect the Sun's rays and keep the Shuttle cool on orbit. The tiles are also treated to protect them from absorbing extra water and adding weight to the craft (figure 7.37).

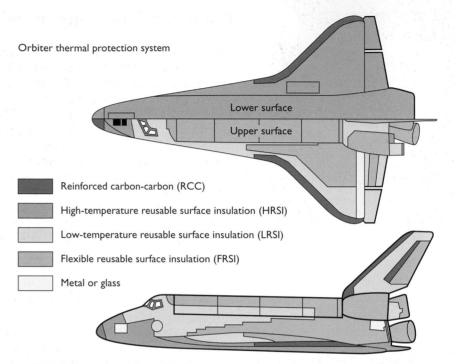

Orbiter thermal protection system

Lower surface

Upper surface

Reinforced carbon-carbon (RCC)

High-temperature reusable surface insulation (HRSI)

Low-temperature reusable surface insulation (LRSI)

Flexible reusable surface insulation (FRSI)

Metal or glass

Figure 7.37 Materials used in the Shuttle.

Example
A piece of space junk has a mass of 50 kg and a speed of 1000 m/s. On re-entry into the Earth's atmosphere the speed decreases to 200 m/s.
(a) Calculate the change in kinetic energy of the junk.
(b) This change in kinetic energy is changed into heat.
    If the specific heat capacity of the material of the junk is 1050 J/kg°C, calculate the change in temperature of the shield.

Solution

(a)
$$E_k \text{ initially} = \frac{1}{2}mu^2$$
$$= \frac{1}{2} \times 50 \times (1000)^2$$
$$= 2.5 \times 10^7 \text{ J}$$
$$= 25\,000\,000 \text{ J}$$

$$E_k \text{ finally} = \frac{1}{2}mv^2$$
$$= \frac{1}{2} \times 50 \times (200)^2$$
$$= 1 \times 10^6 \text{ J}$$
$$= 1\,000\,000 \text{ J}$$

Change in kinetic energy = 24 000 000 J = 2.4×10^7 J.

(b)
$$E_h = mc\Delta T$$
$$2.4 \times 10^7 = mc\Delta T$$
$$2.4 \times 10^7 = 50 \times 1050 \times \Delta T$$
$$\Delta T = 460°C$$

Figure 7.38

Living and working in space

Space is a not a pleasant environment. There is no air, extreme cold when not facing the Sun and unpredictable blasts of radiation from the Sun. Some effects are not so obvious. The weakening of the bone structures can occur during long-term missions, such as the MIR missions which could last up to a year for some of the cosmonauts. The time spent in a weightless environment results in little or no support needed for posture.

The training for space walks involves training in a water environment to simulate weightlessness. But water resists motion and there is nothing in space. One astronaut said: 'You learn to bring yourself to a stop and then make yourself motionless.' The space suit also restricts movement. Despite having moveable parts, the shoulder joint rotates from front to back and not side to side. To lift something you rotate your body until you are facing it.

The gloves are very bulky and the tools are much larger than normal to allow for the lack of sensitivity (figure 7.38).

The International Space Station

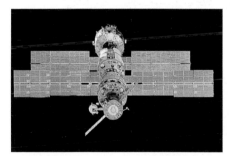

Figure 7.39 The International Space Station.

This is a station built as a series of units, called modules which are being assembled above the Earth. It is over 100 m long and a similar length wide. It has a large area of solar panels. It has been largely assembled by the USA and Russia, although 16 countries are involved in the construction (figure 7.39).

The station will conduct different types of experiments to see the effects of space on a variety of situations. These will include:
◆ Growing tissue in a reduced gravity environment.
◆ Reduced gravity effect on the human body.
◆ Flames, fluids and metals in space.
◆ The nature of space. This will examine how the basic forces of nature will affect materials.

Space forms both a beginning of an adventure for humankind and the end of this book. This is the last part of the course to be developed.

'Mr President, the eagle has landed'.
Note left on President John F Kennedy's grave on 20 July 1969

'It is difficult to say what is impossible, for the dream of yesterday is the hope of today and the reality of tomorrow'.
Robert Goddard, space pioneer

Triumph and Tragedy

Despite all the success of space missions, problems can occur. In 1986, one shuttle, Challenger, exploded shortly after launch, losing all of the astronauts (including a teacher). In 2003, the Columbia exploded on re-entering the Earth's atmosphere.

- ◆ A rocket is pushed forward because the 'propellant' is pushed back.
- ◆ Simple situations like rocket motion can be explained using the rule: A pushes B, B pushes A back.
- ◆ Newton's Third Law is: 'If A exerts a force on B, B exerts an equal but opposite force on A'.
- ◆ Thrust = mass × acceleration, and is an example of $F = ma$.
- ◆ Rocket motors need not be kept on during interplanetary flight.
- ◆ The force of gravity near the Earth's surface gives all objects the same acceleration (if the effects of air resistance are negligible).
- ◆ The weight of an object on the Moon or on different planets is different from its weight on Earth.
- ◆ Objects in free fall appear weightless.
- ◆ The weight of a body decreases as its distance from the Earth increases.
- ◆ Projectile motion can be treated as two independent motions.
- ◆ Satellite motion is an extension of projectile motion.
- ◆ An effect of friction is the transformation of E_k into heat.
- ◆ Numerical questions can be solved using $E_h = cm\Delta T$, work = Fd, and $E_k = \frac{1}{2}mv^2$.

End of Section Questions

1 A space probe to Mars has a mass of 75 kg. What is the weight of the probe:

    (a) on Earth?
    (b) on Mars?

2 (a) Draw a labelled diagram of a rocket on take off.
    (b) Use the idea of Newton's Third Law of motion to explain how a rocket moves in space.

3 In a Shuttle mission, the Shuttle is in orbit around the Earth and is under the influence of the Earth's gravitational field. Astronauts inside the Shuttle are weightless. Explain why this contradiction happens.

4 A rocket has a mass of 1250 kg on take off. The thrust developed by the engines is 15000 N.

    (a) Calculate the weight of the rocket.
    (b) Calculate the unbalanced force acting on the rocket.
    (c) Calculate the acceleration of the rocket.

5 A spacecraft of mass 800 kg is travelling at 350 m/s.

    (a) Calculate the kinetic energy of the spacecraft.
    (b) As it passes through the atmosphere its speed decreases to 150 m/s. Calculate the change in kinetic energy.
    (c) Calculate the change in temperature of the spacecraft if the specific heat capacity of the spacecraft material is 1200 J/kg°C.

Exam Questions

1 A satellite is launched into orbit. It has a detector to observe infrared radiation.

    (a) Name a detector of infrared radiation.
    (b) The wavelength of the infrared radiation is 10.6 × 10^{-6} m. Calculate the frequency of this radiation.

2 During a space mission to assemble the space station, the astronauts have a space walk.

    (a) The mass of one astronaut is 70 kg on earth. Explain if this will change during the space walk.
    (b) What is his weight on Earth?
    (c) Explain why he is weightless during the space walk yet gravity is still present.
    (d) What is meant by gravitational field strength on Earth = 10 N/kg?

3 A space probe lands on the moon to take rock samples. It has a mass of 450 kg.

    (a) What is the weight of the probe on the Moon?
    (b) The thrust exerted by the probe rockets is 1620 N. What is the unbalanced force exerted on the probe?
    (c) Calculate the acceleration of the probe.
    (d) Explain why the acceleration decreases as the probe rises away from the surface.

4 A section of the shuttle has a mass of 300 kg and experiences a temperature change. The specific heat capacity of this part of the shuttle is 1400 J/kg°C.

    (a) The Shuttle slows down from 7500 m/s to 150 m/s. Calculate the change in kinetic energy.
    (b) Only 10% of this energy is converted into heat. Calculate the change in temperature of this part of the shuttle.

REVISION GUIDELINES FOR TESTS AND EXAMS IN PHYSICS

This is a series of advice tips for exams and tests in physics.

◆ Read the learning outcomes for each unit and use this as a checklist. This is the list on which the examiners will base the test or exam.

◆ Pay attention to any descriptions of experiments.
Examples include:
  – determination of half life in unit 3;
  – determination of focal length of a lens in unit 3;
  – measurement of speed and acceleration in unit 5;

◆ Read the question carefully. If you are asked for one reason, do not give two. If your second answer is incorrect it will cancel out your first, possibly correct answer.

◆ Some questions will seem unfamiliar. They are designed for you to use your knowledge of physics in an unknown situation. Think of the most suitable unit from your revision and then decide which information you can use.

◆ In revision, always split the work into small units and revise several subjects over a time given period. This means that you will return to the topics on a shorter time scale, rather than one night on each subject. This means that the memory gain is better.

◆ Revise equations on a regular basis for a very short time scale, for example, 10 minutes per day. After a few days the equations will be remembered easily. Do not try to remember them the night before the test or exam.

◆ If you use the triangle method to rearrange equations, you must still write down the equals sign. The triangle is not an equation.

◆ In solving numerical questions the procedure is:
  – write down the equation;
  – if you need a piece of data check the data sheet;
  – substitute the information;
  – calculate the answer;
  – insert and check the units, for example speed is m/s and acceleration m/s^2.

FORMULAE

Given below are all the formulae from the Standard Grade Physics course.

TELECOMMUNICATIONS.

$v = \dfrac{d}{t}$

$v = f\lambda$

USING ELECTRICITY.

credit $\quad Q = I\,t$

$V = I\,R$

$P = I\,V$

credit $\quad P = I^2\,R$

$E = P\,t$

credit $\quad R_{series} = R_1 + R_2 + R_3$

credit $\quad \dfrac{1}{R_{parallel}} = \dfrac{1}{R_1} + \dfrac{1}{R_2} + \dfrac{1}{R_3}$

HEALTH PHYSICS.

credit $\quad P = \dfrac{1}{f}$

credit $\quad f = \dfrac{1}{P}$

ELECTRONICS.

credit $\quad \text{power} = \dfrac{V^2}{R}$

$\text{voltage gain} = \dfrac{V_{out}}{V_{in}}$

credit $\quad \text{power gain} = \dfrac{P_{out}}{P_{in}}$

TRANSPORT.

$\text{average } v = \dfrac{d}{t}$

credit $\quad a = \dfrac{v - u}{t}$

$W = mg$

$F = ma$

$\text{work} = F \times d$

$P = \dfrac{E}{t}$

$E_p = mgh$

credit $\quad E_k = \tfrac{1}{2}mv^2$

ENERGY MATTERS.

credit $\quad \%\text{ efficiency} = \dfrac{\text{energy out}}{\text{energy in}} \times 100$

credit $\quad \%\text{ efficiency} = \dfrac{\text{power out}}{\text{power in}} \times 100$

$\dfrac{n_p}{n_s} = \dfrac{V_p}{V_s}$

credit $\quad V_p \times I_p = V_s \times I_s$

credit $\quad \dfrac{n_p}{n_s} = \dfrac{I_s}{I_p}$

$E = ItV$

$E_h = cm\Delta T$

credit $\quad E_h = ml$

Index

Answers

Chapter 1 Telecommunications

Section 1.1 Communication using waves

1 0.66 m

2 17 000 Hz

3 (a) 0.5 m (b) 3 cm (c) 3 Hz (d) 0.09 m/s

4 (a) Light travels faster than sound

  (b) 1360 m

Section 1.2 Communication using cables

1 (a) Precise information sent over long distance

  (b) People may not answer

2 (a) Transmitter is microphone; receiver is loudspeaker

  (b) Transmitter: sound to electrical. Receiver: electrical to sound

3 (a) Thin glass solid tube

  (b) Less interference

  (c) See figure 1.12

Section 1.3 Radio and Television

1 (a) Demodulator, loudspeaker

  (b) Demodulator: removes carrier wave. Loudspeaker; changes electrical signal to sound

2 (a) Frequency modulation (b) 3.24 m (c) VHF

3 (a) Red, blue and green

  (b) Different amounts of the basic colours

4 An electron beam sweeping across the screen produces each line. Different intensities are produced by different amounts of electrons striking the screen. A sequence of lines (625) builds up a single picture since the beam is swept back up to the top at the end of the frame

5 15 frames per second is just on the limit of movement with the brain almost seeing this set of pictures as movement. Each frame or picture is slightly different from the last one and the set makes a moving picture but is slightly jerky

6 (a) This allows a low frequency audio wave to be combined with a high frequency carrier wave. The combined wave can travel large distances

  (b) The high frequency carrier wave is removed from the audio wave

Section 1.4 Transmission of radio waves

1 (a) Advantage: communicate anywhere; Disadvantage: battery needs charging

  (b) Signals are absorbed by buildings

2 See figure 1.38 (b)

3 To avoid interference between signals

4 Greater height and longer orbit time

5 Higher frequency means lower wavelength and this means less diffraction

Chapter 1 Exam Questions

1 (a) Modulation

  (b) Number of waves per second

  (c) 300 000 000 = 1515 198 000

2 (a) 8 m/s (b) 1.33 m

3 Red, green and blue

  (b) An electron beam sweeping across the screen produces each line

   Different intensities are produced by different amounts of electrons striking the screen. A sequence of lines (625) builds up a single picture since the beam is swept back up to the top at the end of the frame

  (c) Different frames are all slightly different and the eye perceives this as a moving picture

4 (a) Selects the station

  (b) Increases the amplitude of the signal

  (c) Separates the audio from the carrier wave

5 (a) 0.16 m (b) Diffraction

  (c) Lower frequency means greater wavelength and the greater the wavelength the more diffraction since diffraction increases with wavelength

6 (a) Geostationary satellites orbit at a greater height

  (b) 24 hours to rotate and stays above the same point on the Earth's surface

  (c) 0.24 s

7 (a) Thin solid glass tube (b) 1.5 ms

Chapter 2 Using electricity

Section 2.1 From the wall socket

1 (a) Electric fire – electrical to heat

   (b) Electric food mixer – electrical to kinetic

   (c) Radio – electrical to sound

   (d) Electric iron – electrical to heat

2 (a) Electric drill and hi-fi (b) kettle and refrigerator

3 (a) 3 A (b) 13 A (c) 3 A (d) 13 A

4 (a) 0.50 mm^2 (b) 1.00 mm^2

5 Socket could become wet and conduct electricity through you when touched

6 When the casing of the appliance becomes live a large current passes through the earth wire. The fuse wire melts (blows) so breaking the electrical circuit and the appliance is no longer live and so safe to touch

Section 2.2 Alternating and direct current

1 37.5 C

2 1.2 A

3 2600 s

4 See figure 2.9

5 (a) Charges move in only one direction

   (b) 1.5 J of energy are given to each coulomb of charge passing through battery

6 (a) 230 V 50 Hz

   (b) Charges move in one direction, then in the other direction and back again, i.e. to and fro

7 Peak value is approximately 14 V so student C is correct

8 (a) 230 V (b) 50 Hz (c) 13 A

   (d) Three – the vacuum cleaner is not double insulated, therefore it requires the earth wire in addition to the live and neutral wires

Section 2.3 Resistance

1 See figures 2.13 and 2.14

2 9 V

3 12 Ω

4 0.26 A

5 1058 W

6 1.2 A

7 12 V

8 920 W

9 51 Ω

10 2 A

11 (a) 18 W (b) 3240 J

12 1058 W

13 2.88 Ω

Section 2.4 Useful circuits

1 (a) 0.5 A, 6 V (b) 4 A, 5 V (c) 2.1 A, 8 V

2 (a) 80 Ω (b) 65 Ω

3 (a) 8 Ω (b) 12 Ω

4 (a) 97 Ω (b) 22 Ω

5 (a) 24 Ω (b) 0.25 A

6 (a) 20 Ω (b) 0.5 A

7 (a) 12 V (b) 2 A (c) 2 Ω

8 (a) Connect ohmmeter across each lamp (positions AB, BC, CD and DE) in turn. Working lamps will give a resistance reading on the ohmmeter

   (b) Broken lamp will show an infinite reading on the ohmmeter i.e. an open circuit

9 (a) 60 Ω (b) 3.83 A (c) 115 V

10 (a) 20 Ω (b) 0.5 A

11 (a) 40 Ω (b) 5.75 A

12 (a) 40 Ω (b) 0.3 A (c) 7.2 V

13 10 Ω

Section 2.5 Behind the wall

1 Lighting circuit has a lower value of fuse; uses thinner cable; is not connected in a ring

2 Ring circuit uses thicker cables

3 3 kWh

4 1 kWh

5 (a) 12 A (b) 6.0 A (c) Circuit B

Section 2.6 Movement from electricity

1 Washing machine; hi-fi; video recorder

2 When switch S is closed it completes the electrical circuit and a current passes through the coil of the electromagnet. The magnetic field produced attracts the iron on the pivoted arm and this pushes the contacts on the other circuit closed. This completes the electrical circuit and so lamp X lights

3 (a) Brushes = E (b) Commutator = D

   (c) Field coils = C (d) Axle = A (e) Rotating coils = B

4 (a) (i) Increases force on coil so coil turns faster
   (ii) Coil turns anti-clockwise
   (iii) Coil turns anti-clockwise

   (b) The commutator automatically reverses the current through the rotating coil every half revolution

5 Good conductor; moulds itself to shape of commutator; withstands high temperature; reduces wear on commutator – any three

Chapter 2 Exam Questions

1 (a) (i) Z
   (ii) Requires 13 A flex as power rating is greater than 700 W; requires a three-core flex since vacuum cleaner is not double insulated

   (b) 38 Ω

2 Large current drawn from socket could overheat socket and cause fire

3 (a) 230 V **(b)** 0.065 A **(c)** 78 C

   (d) Charge alternates by moving in one direction and then in the opposite direction, i.e. to and fro

4 (a) Current **(b)** 53 Ω

5 (a) Resistance

   (b) Working lamps will give a resistance reading on the ohmmeter. Broken lamp will show an infinite reading on the ohmmeter

   (c) An open circuit

6 (a) Circuit A

   (b) (i) 200 Ω **(ii)** 50 Ω

   (c) (i) 1.15 A **(ii)** 4.6 A

   (d) Circuit B – it has the higher power rating since lower resistance means higher current reading for the same voltage

7 (a) (i) 15 Ω
   (ii) Current = 0.67 A so ammeter is not correct

   (b) (i) 1.5 Ω **(ii)** 6.7 A

8 (a) X = brown; Y = blue; Z = green/yellow

   (b) (i) 5.4 A **(ii)** 32.5 p

9 (a) A = field coils (magnet), B = commutator, C = brushes, D = rotating coil

   (b) Reverse direction of magnetic field; reverse connections to battery

Chapter 3 Health Physics

Section 3.1 Body temperature

1 Small temperature range allows an accurate reading to be taken; a kink to stop liquid flowing back into bulb

2 If temperature is above or below 37°C, then patient is ill and further tests are needed

3 34°C is below normal; there is a danger of hypothermia

Section 3.2 Sound and ultrasound

1 Open bell for the heart and closed bell for the lung

2 (a) Frequencies greater than 20 000 Hz

   (b) Breaking up of kidney stones; to look at foetus in womb

3 0.002 m

4 (a) 90 dB **(b)** Dullness of hearing

Section 3.3 Light and sight

1 To receive light signals

2 Smaller and upside down

3 Bends towards the normal entering the glass and away from the normal on leaving the block

4 (a) This is the distance from the lens to the focal point

   (b) Using an object such as a window frame, the outline of the frame is focussed on a card. The distance from the lens to the card is measured with a ruler. This is the focal length

5 (a) Short sight **(b)** Concave

6 8 D

7 Cold light source

Section 3.4 Using the spectrum

1 Sealing detached retinas; removing tattoos

2 Infrared

3 (a) Vitamin D **(b)** Skin cancer

4 (a) Photographic plate

   (b) Black line since radiation passes through break

5 Three-dimensional image or more detail shown

Section 3.5 The atom and radiation

1 A=alpha, B=gamma

2 (a) Gain or loss of electron producing a charged particle

   (b) Large number of ions in a small volume

3 (a) Becquerel (Bq)

   (b) Sievert (Sv)

   (c) Nature of radiation; type of body tissue, energy of radiation

4 Gamma passes through tissue and can be detected outside body, others cannot

5 Do not handle source. Point away from the body

6 Radioactive energy being changed into light energy

7 (a) Time taken for the activity of the source to reach half of its initial value

(b) Almost 7 hours

8 (a) Radiation from natural sources

(b) 200 counts per second

Chapter 3 Exam questions

1 (a) There is a kink to stop the liquid running back to the bulb; smaller range over which changes in the volume of the liquid take place, therefore the thermometer is accurate

(b) 34°C is lower than normal temperature, this could cause hypothermia

(c) (i) Photographic film

(ii) Shows image in three dimensions

2 (a) Distant objects are blurred and near objects are in focus

(b) Concave **(c)** −6.67 D

3 (a) Frequencies greater than 20 000 Hz

(b) 3×10^{-4} m **(c)** Less absorption

4 (a) Skin manufactures Vitamin D when exposed to ultraviolet light

(b) Skin cancer

(c) (i) Infrared **(ii)** Longer wavelength

5 (a) Passes through tissue

(b) Too short for radiation to reach the organ

(c) 3.25 kBq

6 (a) Thin glass solid tube

(b) Cold light source

(c) See figure 1.12

Chapter 4 Electronics

Section 4.1 Electrical Systems

1 (a) Digital **(b)** Analogue **(c)** Analogue **(d)** Digital

2 Analogue – cassette recorder; radio; television; thermometer (mercury)

Digital – CD player; computer; thermometer (electronic)

Section 4.2 Output Devices

1 (a) Electrical to kinetic

(b) Electrical to light

(c) Electrical to sound

2 (a)

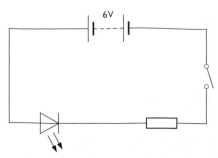

A2.1

(b) Resistor is required to prevent the LED being damaged by too large a current passing through it (or too large a voltage across it)

3 600 Ω

4 (a) 2 = a, b, g, e, d **(b)** 5 = a, f, g, c, d **(c)** 7 = a, b, c

5 (a) buzzer or lamp **(b)** motor **(c)** relay

6 (a) 2 = 0 0 1 0 **(b)** 5 = 0 1 0 1

(c) 7 = 0 1 1 1 **(d)** 8 = 1 0 0 0

7 (a) 0 0 0 1 = 1 **(b)** 0 1 0 0 = 4

(c) 0 1 1 0 = 6 **(d)** 1 0 0 1 = 9

Section 4.3 Input devices

1 (a) Sound to electrical **(b)** Heat to electrical

(c) Light to electrical

2 (a) Thermocouple or thermistor **(b)** LDR or solar cell

(c) Capacitor **(d)** Microphone

3 (a) 40 Ω **(b)** 0.15 A **(c)** 1.5 V

4 (a) $V1 = 1$ V, $V_2 = 5$ V

(b) $V_1 = 8$ V, $V_2 = 4$ V **(c)** $V_1 = 8$ V **(d)** $V_2 = 3$ V

5 (a) 500 Ω

(b) Resistance of thermistor decreases so ammeter reading will increase to, for example, 0.0040 A

6 As light intensity increases, the resistance of the LDR decreases. Voltage across LDR will decrease and so the voltage across the resistor will increase and so voltmeter reading increases

7 (a) Increases

(b) Increase value of resistor or increase value of capacitor

Section 4.4 Digital processes

1 (a) X = thermistor, Y = transistor; Z = LED

(b) Electronic switch

2 See Section summary

3

A	B	C	D
0	0	1	0
0	1	1	1
1	0	0	0
1	1	0	0

4

A	B	C	D
0	0	1	1
0	1	0	0
1	0	1	1
1	1	0	1

5 Logic gates are NOT and AND

6 As the light level on the LDR increases, the resistance of the LDR decreases and so the voltage across the LDR decreases. The voltage across the resistor must therefore increase and when it is equal to or greater than 0.7 V the transistor switches on and so the LED lights

7 Frequency is reduced

Section 4.5 Analogue processes

1 Hi-fi, radio, and television

2 (a) To make the output signal larger than the input signal

   (b) Same frequency

3 500

4 5.5 V

5 0.033 W

6 (a) 9.8×10^{-9} W (b) 1.53×10^{9}

Chapter 4 Exam Questions

1 (a) NOT gate and transistor

   (b) Light to electrical

   (c) Capacitor (d) 0.0056 A

2 (a) (i) 4 V (ii) 0.004 A (iii) 1250 Ω

   (b) As the light intensity increases, the resistance of the LDR decreases and so the voltage across the LDR decreases. The voltage across the resistor must therefore increase

3 (a) X = thermistor; Y = transistor; Z = LED

   (b) As the temperature of the thermistor increases, the resistance of the thermistor decreases and so the voltage across the thermistor decreases. The voltage across the variable resistor must therefore increase and when it is equal to or greater than 0.7 V the transistor switches on and the LED lights

   (c) Change the positions of the thermistor and variable resistor

4 (a)   (b) 180 Ω

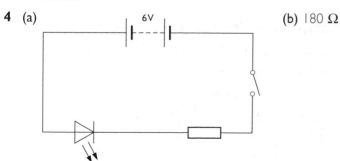

EA4.1

5 (a) 4.5 V

   (b) (i) Transistor

      (ii) As it gets darker, the resistance of the LDR increases and so the voltage across the LDR increases. The transistor switches on and there is a current in the relay. The relay switch closes and the lamp lights

   (iii) 0.26 A

6 (a) Capacitor charged, logic level at P = 1, logic level at Q = 0

   Capacitor uncharged, logic level at P = 0, logic level at Q = 1

   (b) (i)   (ii) logic 0

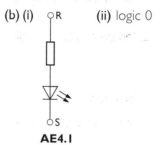

AE4.1

   (c) Increase resistance of resistor or increase capacitance of capacitor

7 (a) 11.4 V (b) 30

Chapter 5 Transport

Section 5.1 On the Move

1 (a) 20 m/s (b) 6 s

2 Use light gate and timer; measure the length of the bicycle and the time to break the light beam

   Speed = length of bicycle / time to break light beam

3 Every second the speed increases by 2.5 m/s

4 2 m/s²

5 (a) −1 m/s² (b) 100 m

6 (a) 0.67 m/s² (b) 48 m/s (c) 8 s

7 (a) (i) 1.25m/s² (ii) 1.33m/s² (iii) −2 m/s² (iv) −1 m/s²

   (b) (i) 10 m (ii) 9.75 m (iii) 9 m (iv) 40 m

Section 5.2 Forces

1 2 m/s²

2 30 N

3 (a) −2 m/s² (b) 1600 N

4 5000 N

5 (a) 500 N (b) 80 N

6 Seat belt exerts a force in the opposite direction causing a deceleration

7 The gaps produce a frictional force

Section 5.3 Work and energy

1 (a) 30 000 J (b) 8 N (c) 6 m
2 (a) 700 J (b) 11.67 W
3 1500 J
4 4 m
5 33.3 W
6 18.75 J
7 50 kg
8 18 J
9 (a) 3.6 J (b) 1.44 m

Chapter 5 Exam Questions

1 (a) a Use light gate and timer

   b Measure the length of the car and the time to break the light beam

   c Speed = length of car / time to break light beam

   (b) 10.5 m/s
2 (a) 2.5 m/s^2 (b) 1.4 m/s^2, Model A (c) 102 400 J
3 (a) Forces are equal (b) 960 J
4 (a) 1500 000 N (b) Streamlined
5 (a) (i) 21 N (ii) 26 N

   (b) 175 J
6 (a) 7200 J (b) 2400 W

   (c) Energy changed to heat and sound

Chapter 6 Energy matters

Section 6.1 Supply and demand

1 (a) Coal, oil and gas

   (b) They are the chemical remains of plants and animals that lived millions of years ago
2 (a) Fit loft insulation; fit draught excluders; etc.

   (b) Use public transport; walk; etc
3 Renewable = biomass; solar; water; wind;
   Non-renewable = coal; gas; oil
4 Renewable sources of energy are sources that can be used again and again. Non-renewable sources of energy are forms of energy that are being used up, i.e. finite, such as the fossil fuels
5 Wind – clean source but unreliable due to variations in wind speed. Solar – clean source but only available during daylight hours. Water – clean source but unreliable due to variations in rainfall

6 (a) On top of a windy hill

   (b) 45 MW

   (c) 700 000 kWh
7 2000 m

Section 6.2 Generation of electricity

1 33.3%
2 (a) 5.25 × 10^5 J (b) 1.05 × 10^5 W
3 33.3 W
4 47.8%

Section 6.3 Source to consumer

1 Move wires faster through the magnetic field; increase number of turns of wire that pass through the magnetic field; increase the strength of the magnetic field of the magnet
2 X = stator; Y = rotor (field coils); Z = dynamo (exciter)
3 (a) a.c.

   (b) When the d.c. supply is switched on the lamp very briefly lights and then goes out. On switching on, the current is initially changing in value in the primary coil. This gives a changing magnetic field which gives a voltage across the secondary coil and so across the lamp initially lights. However, when the current is constant through the primary coil, this gives a constant magnetic field and so no voltage is produced across the secondary coil and so the lamp is unlit
4 (a) 180 turns (b) 23 V

   (c) 40 V; 8 A (d) 1.25 V; 800 turns
5 (a) 115 V (b) 4 A (c) 2 A
6 Any two from: heating in coils; heating of the magnetic core; sound; all of the magnetic field from the primary does not pass through the secondary
7 (a) 0.4 A (b) 3.2 W (b) 4 W (d) 80%
8 (a) 2 A (b) 20 W

Section 6.4 Heat in the home

1 (a) e.g. double glazing reduces conduction from windows due to air gap between panes being a poor conductor of heat and cavity wall insulation would reduce conduction through walls due to the insulation being a poor conductor of heat

   (b) e.g. loft insulation and cavity wall insulation would both reduce convection by trapping a layer of air

   (c) e.g. putting silver foil behind a radiator reduces radiation losses as heat from the radiator is reflected back into the room; putting silver foil above the ceiling reduces radiation losses as heat radiation from the room below is reflected back into the room; a layer of gold on windows allows radiation from outside the house to pass inside but heat radiation from the room is reflected back into the room

2 (a) 10 000 J (b) 15 000 J (c) 20 000 J (d) 100 000 J

3 500 J/kg°C means that 500 J of energy are required to change the temperature of 1 kg of the glass by 1°C

4 83 600J

5 14 000 J

6 194 s

7 (a) 17 280 J (b) 19°C

8 668 000 J

9 4.52×10^6 J

10 0.088 kg

Chapter 6 Exam Questions

1 (a) Hydroelectric, nuclear or wind

(b) Any one from solar, wave, tidal, geothermal or biomass

(c) Unreliable

2 (a) 9 J (b) 48 J (c) 18.75%

(d) Some of the electrical energy supplied to the motor is converted into heat and sound energies. This means that the useful energy output i.e. the gain in gravitational potential energy is less than the electrical energy supplied and so the efficiency is less than 100%

3 (a) a.c. voltmeter

(b) Rotating spindle causes the magnet to rotate. The magnetic field through the coil of wire changes and so a voltage is produced. This voltage causes the pointer on the voltmeter to be deflected

(c) The magnet rotates faster so the magnetic field through the coil is changing more rapidly and a larger voltage is produced. Hence larger reading on voltmeter

(d) Use a magnet with a stronger magnetic field and put more turns of wire on coil

4 (a) 12 V (b) 4 A (c) 0.21 A

5 (a) 6 A

(b) (i) Step-down transformer

(ii) 97%

(iii) Any one from heating of the coils, heating of the iron core, sound, or leakage of magnetic field from the transformer

6 (a) 62 700 J

(b) 45.4 W, that all the energy supplied by the heater is absorbed by the water

(c) 3.78 A

7 (a) 80°C (b) 1375 J/kg °C (c) 150 000 J/kg

Chapter 7 Space physics

Section 7.1 Space – the final frontier?

1 See figure 7.9

2 Probe must detect infrared as well as light

3 (a) See figure 7.10

(b) Image is larger and virtual

4 (a) Fluorescent film (b) phototransistor

5 Brings parallel rays to a focus

6 (a) Temperature and the elements in the star

(b) Y17 is hotter

7 (a) November 1973 (b) 100 000 years

Section 7.2 Jets and rockets

1 (a) 750 N (b) 300 N

2 (a) Upwards arrow labelled thrust; downwards arrow labelled weight

(b) Hot gases are produced and forced backwards this is the action forces and the reaction force is rocket moving forwards

3 Shuttle and astronauts are falling at the same rate

4 (a) 12500N (b) 2500 N (c) 2 m/s²

5 (a) 4.9×10^7 J (b) 4.0×10^7 J (c) 41.6°C

Chapter 7 Exam Questions

1 (a) Phototransistor

(b) 2.8×10^{13} Hz

2 (a) No change since mass is unaffected since not changed by gravitational forces

(b) 700 N

(c) The astronaut is falling at the same rate as the spacecraft

(d) The gravitational force on 1 kg is 10 N

3 (a) 720 N (b) 900 N (c) 2 m/s²

(d) Since fuel is being used up mass is decreasing

4 (a) 8.44×10^9 J (b) 1200°C